Raising Healthy Animals Series

Every year, thousands of people around the world struggle to survive because they don't have the right knowledge, skills and resources to care for their animals. Christian Veterinary Mission (CVM) sends veterinary professionals to live and work alongside many of these people to encourage them and provide them with not only much needed veterinary expertise, but also the hope that is only found in Christ. CVM veterinarians build lasting relationships with individuals and communities, helping them be transformed through Christ's love.

CVM, in its effort to be meaningfully involved in work in the developing world, quickly found there was little appropriate educational material available. CVM set about developing basic resource materials in animal husbandry for farmers and agricultural workers. Apparently, they met a real need, as these books have been accepted throughout the developing nations of the world.

The series of books published by Christian Veterinary Mission includes the following in order of publication:

Raising Healthy Pigs *	Drugs and Their Usage
Raising Healthy Rabbits *	Where There Is No Animal Doctor
Raising Healthy Fish	Raising Healthy Horses
Raising Healthy Cattle	Zoonoses: Animal Diseases That Affect Humans
Raising Healthy Poultry *+	Raising Healthy Honey Bees
Raising Healthy Goats *	Slaughter and Preservation of Meat
Raising Healthy Sheep	Disease and Parasite Prevention in Farm Animals

[Also available in: * Spanish + French].

CVM fieldworkers have also developed specific training materials for the countries in which they work.

All of these books have been put together by Christian men and women; in a labor of love and service, for people in need throughout the world. It demonstrates dedication to their profession, service to humanity and a witness to their faith. We hope that they are a help to you in developing an appropriate livestock program to meet your needs. We pray God's blessing on their use.

Leroy Dorminy
CVM Founder

TABLE OF CONTENTS

Christian Veterinary Mission (Publisher of this book)

Our vision is to see

Christ's love expressed through veterinary medicine.

Our mission is to

challenge, empower and facilitate veterinarians to serve through their profession, living out their Christian faith.

CVM also provides education and encouragement for those who desire to minister through service, prayer, relationship building, and modeling Christ's love.

About CVM

Christian Veterinary Mission (CVM) is a registered non-profit Christian Service Organization 501(c)(3) based in Seattle, Washington, U.S.A.

CVM was founded in 1976 by Dr. Leroy Dorminy who came to realize the impact that veterinarians could have by integrating their faith with their practice, both locally and around the world. In 2008, CVM had nearly 30 veterinary professionals serving full-time internationally and over 200 veterinary professionals and student volunteers serve on short-term cross-cultural mission trips annually. CVM sponsors fellowship & prayer breakfasts at over 20 U.S. veterinary meetings each year and reaches out to veterinary students through Christian Veterinary Fellow-ship (CVF) groups in every veterinary school in the U.S. by encouraging them in spiritual growth and professional development.

There are over 3,500 veterinarians affiliated with CVM in the U.S. CVM also partners with organizations and networks in other countries that are focused on empowering Christian veterinarians. CVM has a volunteer advisory board of veterinarians who guide its vision, mission, and programming.

CVM books and the free International Animal Health Newsletter were written with small farmers, veterinarians, and agricultural development workers in mind. Our desire is that they would help individuals and groups develop an appropriate livestock program to meet community needs. CVM's Endowment Fund was started in the early years of the organization's life. The fund provides for meaningful programs that could not be funded by the regular budgeting process.

(GROUND

oklet was prepared by Christian Veterinary Mission as an aid to working in areas with limited sources of information and assis- he main purpose is to provide a selected set of common basic s which fish farmers may be able to adopt in order to help eradi- ontinuing husbandry problem.

aquaculture is practiced so widely throughout the world, we give s, where possible, of husbandry practices in use in specific coun- hope this book will provide the reader ideas that will be of benefit. cies used and management practices suggested are directed creasing protein availability in subsistence farming situations. not suggestions of economically profitable ventures.

armer wishing to pursue this venture for the first time, it is recom- begin on a small scale. Once he gains experience he can expand.

)UCTION

over 25,000 species of fish worldwide. Dozens of species are mmercially and many more are caught for human consump- tic animals other than fish such as shrimp, crabs and mus- re farmed. Different aquatic animals have adapted to life in nts with temperatures from below freezing to over 32°C from almost pure rain water to salinities of over 4%. Some spend part of their lives in non-aquatic environments. Pro- tems will therefore vary according to the specific needs of f aquatic species. A major advantage of fish farming is that ion of protein from a given area is potentially much higher attle, pigs, or plant crops. Also, it is possible to blend aqua- existing culture of other animals and crops, a practice known d agriculture-aquaculture. This additional aquatic crop can a cash for a farmer, or provide an extra source of protein.

planned aquaculture operation can allow certain farmers nore self-sufficient. Fish are a high value crop that can be and urban markets. However, aquaculture is much differ- r forms of husbandry, as start up and operating costs can risk is high. Also, a successful operation requires suitable d climate. Therefore, aquaculture is not suited for all farm- of the complexities of the environment, economics and ll levels, it is recommended that anybody willing to try tart with caution at a small, low intensity scale.

l focuses on the techniques for raising healthy fish. Careful development stage is essential for a successful operation. design and species selection are the most critical concerns anning stages. Site selection and design are interrelated d by environmental parameters covered later in this text. In

Section I

Background

Permission requests should be addressed to: vetbooks@cvm.org

OR

Christian Veterinary Mission
www.cvm.org

THIS IS A REVISED THIRD EDITION

We feel it has been improved through this revision. We hope it will be a valuable addition to your efforts at Fish Farming. If you have questions or suggestions concerning the material covered in this book you may write to the author at:

e-mail:

Christian Veterinary Mission
vetbooks@cvm.org

Library of Congress Cataloging-in-Publication Data

Petrie-Hanson, Lora.
Raising healthy fish III / by Lora Petrie-Hanson and Larry Hanson.— Rev. 3rd ed.
p. cm. — (Raising healthy animals series)
Includes bibliographical references (p.).
ISBN 1-886532-05-2
1. Fish-culture. 2. Fishes—Health. I. Hanson, Larry. II. Christian Veterinary Mission. III. Title. IV. Series.
SH151.P38 2004
639.3—dc22

200400129

Raising Healthy Fish

by

Members of the Aquatic Task Force
College of Veterinary Medicine
Mississippi State University
MS 39762 USA

Christian Veterinary Mission

Second and Third editions by
Lora Petrie-Hanson and Larry Hanson

First edition by

Lydia Brown
Julia Heard
J. Randy MacMillan
Tom Nemetz
Andrew Grady
E. Mac Huddleston
A. Doug Munson
Chris Wilson

A Publication of Christian Veterinary Mission

Current Book Information at
www.cvm.org

*Give a man a fish and you feed him for a day.
Teach a man to raise fish and you feed him for life.*

selecting the species to be cultured, economic, ecological, technological and environmental constraints must be considered.

Economic factors are one of the first considerations. In the broadest sense, economic factors can include labor and nutrient input for subsistence operations. For subsistence farms it is usually best to grow fish low on the food chain (such as plant eaters) rather than a carnivorous species because food requirements for the latter are often of better quality and more expensive than what the farmer feeds his own family. On the other hand if the fish are to be sold as an economic supplement, local market acceptability must be evaluated. Factors to consider when evaluating the potential market are species, size and form of product that is to be marketed. A high value item that is difficult to transport to the market may be less desirable than a lower value, lower risk product.

Ecological factors are often overlooked, having potentially disastrous results. In general, native species have the best local market potential and reduce the risks of importing exotic problems. Agriculture is notorious for introducing exotic species throughout the world without regard to the detrimental effects the introduced species may have on the native plant and animal life or current agricultural commodities. The effects on native wildlife, fisheries, forestry or crops have often been more detrimental than the benefit derived from the new agricultural industry. The risk of introducing an exotic species of fish must be evaluated very carefully by individuals with professional expertise in fisheries biology and ecology before the fish are brought into the area. Fish will escape from an aquaculture operation so the ability of the fish to reproduce and survive in the wild must be assessed. Less obvious risks of introducing fish from out of the region include the introduction of new diseases to native populations. Introduced diseases may decimate native fish populations. Additionally, introduced species can become an agricultural pest. Examples of the devastating effects of introduced exotic aquaculture species include:

—The North American red swamp crawfish was introduced into Spain and became a serious pest to the rice crop.
—North American crawfish imported the crawfish plague into Europe and destroyed the natural stocks of crawfish and the industry associated with it.
—Importation of Asian carp imported the Asian tapeworm into North America causing damage to several species of North American cyprinids that are important forage fish for natural populations.
—Importation of Tilapia into Florida resulted in population explosions that impacted natural populations and damaged sport fisheries.

Technological constraints include the basic knowledge of the biology of the fish chosen. Can the fish be reliably spawned? Can reproduction and fish density be controlled? What are the nutrient requirements? Are the fish territorial or cannibalistic?

Environmental constraints are mainly based on water quality, temperature and salinity. Temperate-water fish can often survive cool temperatures but grow poorly, warm water fish may die from even short exposures to temperatures below 15°C, while cold water fish die at temperatures exceeding 22°C. Pond or stagnant water fish can generally survive in water containing lower levels of oxygen than can fish species that live in well aerated streams. Fresh water fish do best in salinities ranging from 0.0001% to 0.5%, brackish water fish do best between 0.5% to 1.5% and marine species do best from 1.5% to 3.5%.

Commonly Cultured Species

Marine species of fish that have been successfully cultured include eels, (*Anguilla* species), flatfish (turbot, *Scophthalmus maximus*), groupers (*Epinephelus* species), mullet (*Mugil* species), pompano (*Trachinotus carolinus*), porgy (black porgy, *Mylio macrocephalus*), puffer (*Fugu* species), salmonids (*Salmo, Oncorhynchus* species), sea bass (*Dicentrarchus* species), sea bream (*Sparus* species), tuna (*Thunnus* species), and yellowtail (*Seriola quinqueradiata*). Early life history is not known for many marine species and captured juvenile natural stocks are often raised to marketable sizes.

Brackish water aquaculture has been performed using cichlids (Tilapia, *Oreochromis* species), flatfish (*Pleuronectes platessa),* groupers (*Epinephelus* species), milkfish (*Chanos chanos*), mullet (*Mugil* species), pompano (*Trachinotus carolinus*), porgy (black porgy, *Mylio* macrozcephalus), puffer (*Fugu* species), sea bass (*Dicentrarchus labrax),* yellowtail (Seriola quinqueradiata), sea bream (*Sparus* species), sea perch (*Lates calcarifer*), and shrimp (*Penaeid* species).

Most small scale aquaculture is performed in freshwater. The species cultured include cichlids (Tilapia, *Oreochromis* species), eels (*Anguilla* species) ayu (*Plecoglossus altivelus*), catfish (channel catfish, *Ictalurus punctatus,* European catfish-weakfish, walking catfish *Clarias batrachus*), cyprinids (common carp, *Cyprinus carpio*), the Chinese carps (grass carp, *Ctenopharyngodon idella,* silver carp, *Hypophthalmichthyes molitrix,* bighead carp, *Aristichthys nobilis*), the Indian carps (catla *Catla catla*, rohu *Labeo rohita* and Mrigal *Cirrhina mrigala*), *Heterotis niloticus*, labyrinth fish (A*nabantidae),* salmonids (*Salmo, Oncorhynchus* species), Nile perch (*Lates niloticus*), Snakeheads (*Ophicephalus* species), striped bass (*Morone saxatilis*), sturgeons (*Acipenser species*), white fish (C*oregonus lavaretus*), and shrimp (M*achrobrachium* species).

Anadromous species leave salt water and ascend rivers into fresh water to spawn. Species, which are shown in more than one grouping, are euryhaline; that is, they are able to withstand widely differing salinities. Catadromous species descend from fresh water into brackish water to spawn. They must remain in brackish water for larval production.

Section II

Water Quality

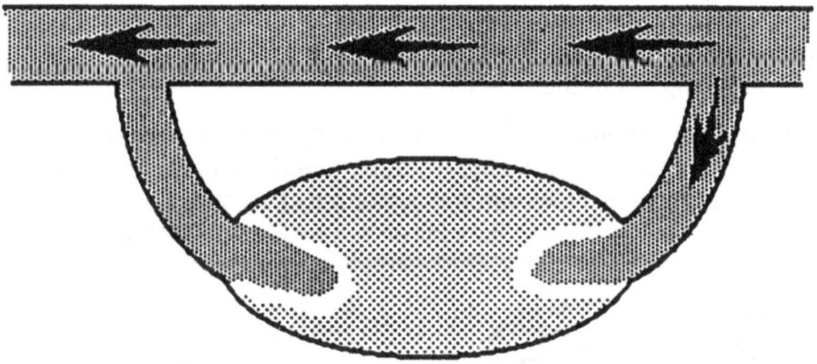

Tidal Flow Seawater

wastes, large populations of predators and wild fish and introduction of pathogens from wild fish.

Water Quality Parameters

Several chemicals are needed in water to assure good growth and survival of fish. Some chemical characteristics are fairly constant. These characteristics are often determined by the mineral characteristics of the soil in the pond, watershed or water source. The constant chemical characteristics of the water can be very important for overall health of the fish being raised and may affect treatment and management strategies. Because they are relatively constant, chemical analysis can be done on an occasional basis. Water can often be sent to a laboratory for analysis. Other chemical parameters are transient and vary substantially as organisms in the water affect the characteristics of the water. Management and monitoring of transient chemical parameters are determined by the intensity of the aquaculture operation. Relatively simple to use test kits are available for farmers to measure the concentrations of these chemicals (Kit manufacturers include Hach Co. and LaMotte). Any manager of an intensive aquaculture system must have the capacity to monitor the critical chemical parameters on a routine basis. Doing so will help to identify water quality problems before they become a serious threat to the fish.

Constant Chemical Parameters:

These parameters are determined by the mineral content in the water. They represent different types of dissolved salts or ions. Because a minimum amount of these salts are critical for the survival of the fish, optimizing the constant parameters are important for maintaining healthy fish.

Salinity is a measure of all ions dissolved in the water. The major component is sodium chloride (the component in table salt). Fish must maintain salt in their body fluids at around 1.8%. Therefore in freshwater (less than 0.5%) fish must expend a considerable amount of energy on maintaining their salt levels and in salt water (3% and over) fish expend energy to expel salt. Marine and freshwater fish are adapted to their environment and the salt content can be optimized for the species to improve growth and reduce stress. Most freshwater fish do best in water with some salinity, with 0.1% being optimum. This level of salt provides optimum conditions for freshwater fish to handle stress due to low temperature, handling, ammonia, and nitrite. Most freshwater sources have salinities much lower than 0.1% and are still suitable for aquaculture but any increase in salinity up to the 0.1% level will generally improve conditions.

Alkalinity is a measurement of the buffering capacity of water or the ability of the water to resist changes in pH. This provides stability to the pond environment. Some salts (mainly bicarbonate) in water act as buffers. They can combine with acid (hydrogen ions) to prevent a change in the pH and so the water remains near neutral or slightly alkaline. Alkalinity can be measured with a test kit. A value of 50–200 ppm is a good range. If the alkalinity is less than 20 ppm, water treatment is necessary (See below).

Hardness is a measure of the calcium and magnesium levels in the water. Hard water has a large amount of calcium and/or magnesium. It is often related to alkalinity because the buffers generally occur naturally as calcium and magnesium salts. Adding some soap to water and trying to make lather is a quick test of hardness. If a good lather can be made, then the water is probably soft. For aquaculture purposes it is best to determine hardness using a test kit or having a sample analyzed by a laboratory. Calcium and magnesium are important minerals required by all living things. High amounts of these minerals are especially important in the water during egg and fry stages, during cold weather, and during stressful conditions. Generally fish grow best in water with high hardness and alkalinity. The minimum hardness for most fresh water aquaculture is 20 ppm and the optimum is over 50 ppm. Even higher levels will help fish that are exposed to very low temperatures or handling stress. Alkalinity and hardness can be increased by adding agricultural limestone ($CaCO_3$) to a pond.

Transient Chemical Parameters:

These parameters are strongly affected by the living organisms in the pond and thus by the day to day management of the farmer. Organisms that constitute the largest mass and thus are the strongest influence on the transient parameters are the microscopic plants (phytoplankton), animals (zooplankton) and bacteria. Factors that affect the growth or cause the death of these organisms will have a profound affect on the pond. Some transient parameters are very critical for fish survival. Lethal pond conditions can occur in just a few hours and the entire pond can die in one day. Because the conditions are transient, a farmer must be able to check for toxic conditions at the pond side and should have a test kit available to monitor a toxic situation. Two biological processes strongly affect the pond chemistry: photosynthesis and respiration.

Photosynthesis is the process that plants use to harvest the energy from the sun for growth. This process takes light and carbon dioxide (CO_2) to make sugar and oxygen (O_2). This is a day-time process in which phytoplankton and submerged aquatic plants cause chemical changes in the water. The degree of change is related to light and the density of the algae. Thus turbidity and fertility are parameters that the farmer influences to affect photosynthesis in the pond. The effect of photosynthesis in a pond is to raise oxygen levels and cause the pH to become more basic (higher) because CO_2 is an acid (it is being used up in photosynthesis).

Respiration is the process in which living things obtain energy from sugars. This occurs in plankton, zooplankton, bacteria, plants and fish. The process takes sugar and O_2 to produce energy and CO_2. This process occurs both night and day but because photosynthesis only occurs during the day time, respiration is most influential at night or during cloudy days and in muddy water. The effect of respiration is to lower oxygen levels and cause the pH to become more acid (lower).

Fish require oxygen in the water at fairly high concentrations in order to survive. **Oxygen is the most critical parameter to manage in an aquaculture operation**. More fish die from lack of oxygen in aquaculture systems than from any infectious disease. Stress from low oxygen predisposes fish to diseases. Oxygen is generally measured in parts per million (ppm) or milligrams per liter (mg/L)(ppm=mg/L).

Oxygen enters water through diffusion or photosynthesis. Any action which causes water to be stirred or moved has the effect of drawing oxygen into it from the surrounding air. The simplest example of this is wind, which moves water and causes ripples. In pond water, there is a constant exchange between the production and consumption of oxygen and carbon dioxide.

Photosynthesis does not take place at night and is reduced during cloudy days when bright sunlight is not available. Respiration occurs all the time and is most rapid at high temperatures. Because of this, oxygen levels in ponds fall at night and rise during the day. Lowest oxygen levels occur just before sunrise and are most critical during periods when the water temperature is high. Shortage of oxygen leads to stress in fish. Conditions that kill plankton or a natural plankton die-off is the most devastating situation for the pond because no oxygen can be produced during the day. Fish which are suffering from lack of oxygen will come to the surface, crowd around the inflow and gasp. Oxygen levels of 4 mg/L (ppm) or more are adequate for good fish health. Many fish species can tolerate levels as low as 1 or 2 ppm for a few hours. But fish showing signs of oxygen stress must get more oxygen quickly. If an oxygen meter is not available, or not practical in a specific location or situation, good management and close observation of the fish to detect the first signs of low oxygen will prevent fish losses. An entire pond of fish can die in 2 to 3 hours and those that survive the stressful condition are more susceptible to diseases. Oxygen levels can be increased by agitating the surface of the water with paddles, paddlewheels, or by adding oxygenated water. If this latter method is used, the inflow pipe should be directed upward to produce a shower effect, increasing the oxygen content of the water. **Any aquaculture operation must have a contingency plan to deal with low oxygen conditions rapidly.**

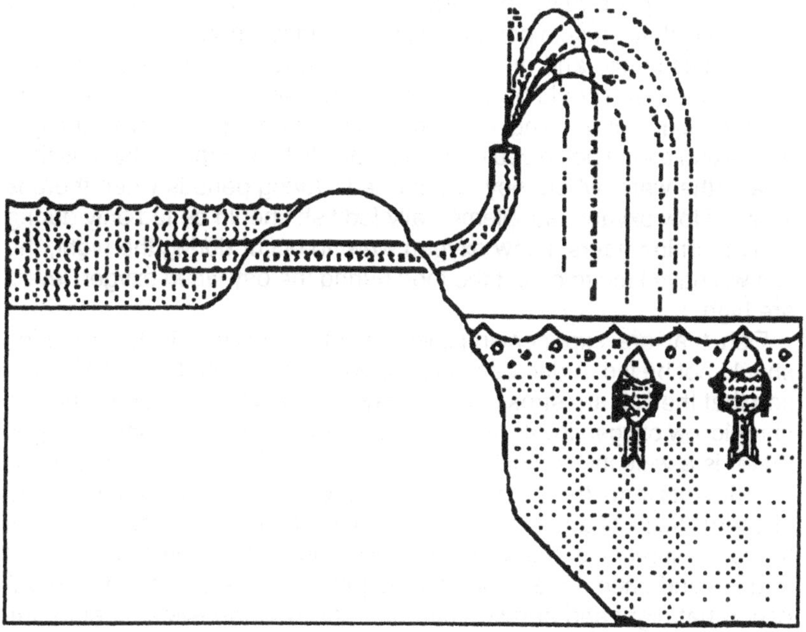

Fish hanging at the surface and gasping for air near the water inflow are classic indicators of a low oxygen situation.

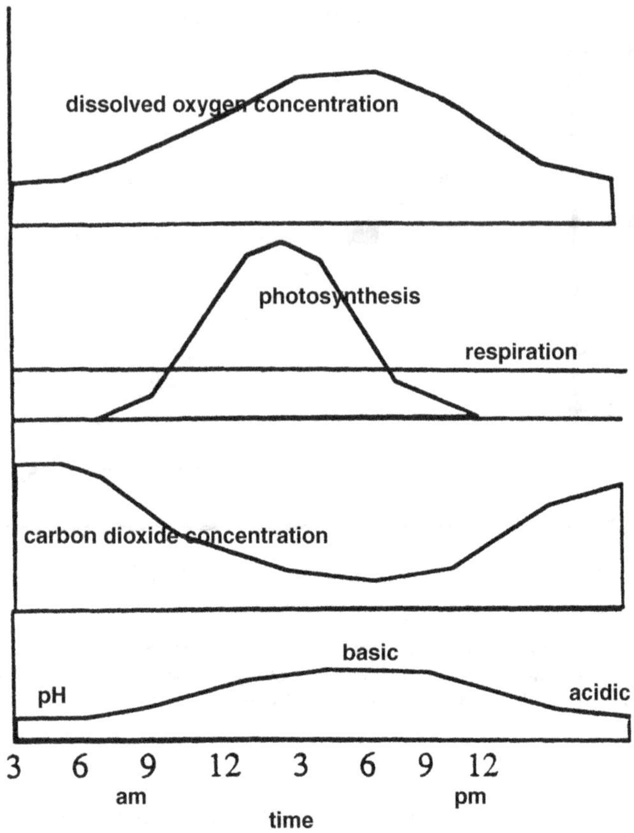

Daily fluctuation in concentration of carbon dioxide, oxygen and pH in ponds associated with photosynthesis.

Turbidity is a measure of the degree of clarity of water. The cloudier the water, the more turbid it is. Turbidity is usually measured by depth of visibility (cm). The less the depth to which it is possible to see through the water, the greater the turbidity. Plankton cause increased turbidity as does suspended clay particles and suspended organic matter. Plankton are often green or bluish. Generally, the greener the color, the more fertile the pond. If pond water has a similar color to the surrounding soil, the turbidity is probably due to particles of soil or silt. In small amounts this is not harmful, but high amounts hinder photosynthesis by blocking sunlight. This limits the productivity of the pond and leads to poor oxygen conditions. Turbidity can be measured with a Secchi disc. Basically a white object is attached to a stick or a string and lowered into the water until the object is no longer visible. This depth is the 'visibility'.

If turbidity due to mud is too high, the cause of the turbidity must be minimized. In hill (watershed) operations, maintaining adequate vegetative cover on the watershed (especially near the pond) is necessary. Also, any livestock should be kept out of the pond and any processes that stir mud from the banks or pond bottom should be minimized.

Ammonia and nitrite Like other animals, fish excrete nitrogenous compounds such as ammonia. In ponds, bacteria convert the excreted ammonia to nitrite and then to nitrate (nitrification). When bacteria are not present in sufficient numbers to degrade the ammonia or if a chemical or environmental insult affects the bacteria, the nitrification process will be inadequate resulting in the build up of ammonia or nitrite. In high concentrations either ammonia or nitrite can rapidly kill fish or stress them (nitrate is not toxic). Ammonia and nitrite are usually measured as ppm-nitrogen. If kits are available, ammonia and nitrite levels can be tested. Fish suffering from nitrite poisoning have brown blood (easily seen as brown gills). These fish usually die very quickly. Levels exceeding 0.75 ppm nitrite-nitrogen can be toxic for some fish species.

Fish suffering from ammonia poisoning may swim in circles or try to jump out of the water. As water temperature and pH rise, ammonia becomes more toxic. Total ammonia concentrations exceeding 3 ppm can be toxic to some species of fish if the pH is over 8.5 at the same time. Treatment for ammonia toxicity is to replace some of the pond water with fresh water. Nitrite poisoning can also be helped in this way. Salt is added to the water to alleviate nitrite toxicity (see section 6). In general, ammonia toxicity is an indication of over stocking and over feeding the fish. This is not a problem in low density operations. Nitrite toxicity can occur following an algae die-off and the rapid degradation of the dead phytoplankton.

Water temperature is important to the growth of fish because they are cold blooded animals. Each species of fish has its own preferred temperature range. Within that range, fish will grow faster at higher temperatures. Extreme fluctuations in temperature always stress fish and lead to reduced growth and possibly even sickness. The oxygen holding capacity of water decreases as the water temperature rises. This is a very important factor when determining stocking rates. There is little that the farmer can do to influence pond water temperature. Pond design can influence rate of temperature change. Deep ponds have lower rates of temperature change than shallow ponds, but will stratify. Stratification is the layering of deep water and occurs because cooler bottom water weighs more than warmer surface water. The cooler bottom water is low in oxygen and when the pond destratifies, the lower portion mixes with the upper portion and causes a severe and sudden decline in oxygen.

Generally ponds are designed to be 1.5–2 meters deep because this allows some depth to buffer rapid temperature changes, allows sufficient depth in moderate turbidity to shade out aquatic weeds and is too shallow for stratification. Another factor influencing temperature is the rapid inflow of cold water from heavy rains which can cool ponds quickly. Ponds with small watersheds are less likely to be affected as drastically by a heavy shower than ponds with large watersheds.

Section III

Facilities

PONDS: The most economically efficient and convenient method of raising fish is in ponds. Some fish can be fed by natural production of food in the pond, which may be increased by adding inorganic or organic fertilizer. Ponds range in size from 0.02 hectare to more than 40 hectares. In general, smaller ponds are easier to harvest, stock, monitor and treat and they involve less individual risk. However, large ponds in general cost less per surface hectare to construct and are more stable i.e., populations of phytoplankton are more stable, rate of temperature change is slower and higher amounts of wind action yield better oxygen characteristics. Ponds can be classified on the basis of their source of water.

1. **Watershed**: Water runs off a large sloping watershed. A sufficient area of watershed must be present to maintain water at adequate levels. The watershed must be protected from erosion and free of agricultural insecticides and herbicides.
2. **Impoundments**: These ponds receive water from streams or springs and have more stable water levels. The incoming water must be screened to prevent introduction of unwanted fish and plants.
3. **Levee Ponds**: These ponds are useful on flat agricultural land. They are formed by construction of a raised wall (levee) around the edge of the pond. The water source is often well water or irrigation water. In other instances a large watershed or impoundment pond will serve as a reservoir for multiple small levee ponds.
4. **Pits or Dug Ponds:** in areas with high ground water and poor water holding soils, ponds can be formed by digging a pit below the water table. These types of ponds are generally small and cannot be drained therefore they are difficult to manage for an aquaculture enterprise.

Soil type is an important consideration in selection of a pond site. Clay has good water holding capacity and is ideal material to use in pond construction. If soil analysis is not available to determine the amount of clay in the earth, one simple test is to squeeze a handful of moistened soil into a ball and drop it from chest height. If the ball does not fragment it should contain enough clay to hold water. Additionally, the soil should be tested for the presence of persistent pesticides if the land was previously used for row crop agriculture. Chlorinated hydrocarbon pesticides such as DDT, toxaphene or chlordane can remain in soils for years. These compounds will accumulate in the fish causing poor growth and survival and can result in a human health risk.

Site preparation time invested in site preparation is well spent. Good planning and development of site in the early stages may help prevent high labor costs later on. It is advisable to remove all trees, stumps, shrubs and large rocks from the pond prior to filling. Any obstruction will make harvesting of fish difficult. Harvesting can be made easier by the construction of a harvesting basin. This is a depression approximately 1.5–2 meters deep and be about 10% of the pond size when the rest of the pond is about one meter deep. When water levels are lowered at harvesting, the fish will congregate in the basin and can be caught more easily.

Dam preparation

The most important phase of building a pond is construction of a solid watertight dam. In order to prevent leakage, the interior "core" of the dam must contain soil with high clay content. The core should be dug down 1 m below the deepest region of non-clay soil and filled in with clay. This should be packed in layers, either by manually tamping the clay or by use of heavy machinery. The outer surface of the dam should be made up of topsoil to promote the growth of grass. This should be fertilized and seeded with grass as soon as possible to prevent erosion. Covering the ground with hay or straw will help prevent erosion until ground cover is established. Trees should not be allowed to grow on the dam since their roots may cause leakage of water.

The dam should have a slope of 3:1 on the front (water) side and 2:1 or 3:1 on the back side for easiest maintenance. A width of 3–5 meters is advised if equipment is to be moved over the dam. Where labor is in short supply, one individual or small family with or without a wheelbarrow can construct a dam one meter wide and one meter deep at its highest point. This would not be suitable for heavy machinery, but would be adequate for a small family farm.

Aquaculture ponds usually do not need to be deep: two meters at the deepest end and 1–1.5 meters at the shallow end is appropriate. Shallower water will encourage the growth of aquatic weeds. Some farmers prefer to have a slightly deeper pond so that the fish can be more sheltered from extremes of temperature and can swim deeper when there is very little other protection from sunlight. However, if the pond is too deep, management and handling of fish is more difficult and the pond may stratify.

Drainage

Some type of drainage is useful to remove excess water from a pond and lower the water level at harvest time. Drainage can be as simple or as complicated as local conditions demand. If working with an existing system that lacks drainage, water can be drained using siphons. A flexible pipe is submerged in the pond, plugged on one end and then the plugged end is dragged out of the pond and over the levee to a point lower than the pond surface and unplugged.

1. **Drain pipe**: A pipe (metal, plastic or bamboo) is placed through a dam during construction and is unplugged to prevent overflow or to drain the pond. Any structure traversing a dam should contain at least two anti-seep collars to prevent leaks from forming around the pipe. **Caution: open drains deeper than 2 meters have considerable vacuum. A swimmer trying to unclog a drain or open a stuck valve can easily be trapped and drown!**

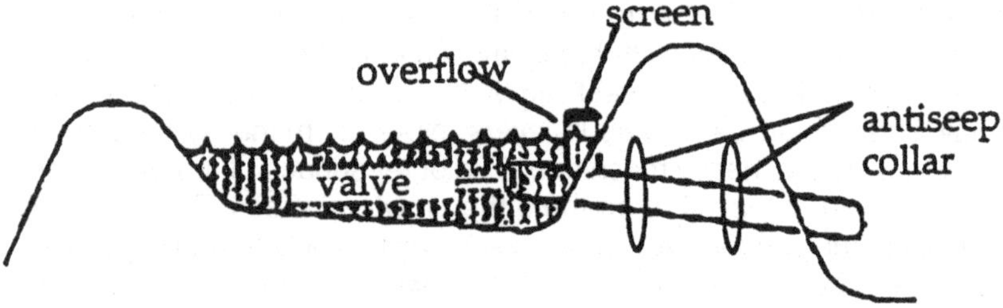

2. **Rivaldi valve**: A flexible plastic pipe is placed through the dam and staked up. It acts as an outflow drain or can be lowered to drain the pond.

rivaldi valve

A 10-cm diameter pipe will be adequate for levee ponds up to 1.2 hectares. A 15–30 cm diameter pipe would be necessary for larger ponds. Hill-shed ponds require at least a 30 cm drain and a spillway system to allow for outflow during times of heavy rain. The spillway should be located at a side of the pond, be very shallow and moderately sloped to direct the water away from the dam, to prevent dam erosion, and covered with rock or a good layer of sod. The broader and shallower a spillway is the less chance of fish swimming out of the pond. The pond should be designed so that the spillway flows only during very unusual situations as wild fish will swim upstream and contaminate the pond.

RACEWAYS

Raceways are linear (rectangular) facilities used for intensive fish culture. Generally, high volumes of water enter at one end and exit at the other end. Raceways are often lined with concrete, brick, gravel or tile. Earthen walls can be used but are subject to erosion. Dimensions of raceways vary but generally follow a length: width: depth ratio of 30:3:1. The main advantage of raceways is the ability of the farmer to raise large quantities of fish in a small area. The main drawbacks to raceways are costs of construction, the need for large water quantities, and the necessity of intensive management.

CAGES AND PENS

Culturing fish in cages and pens is another method of intensive fish culture where fish are placed in wooden, wire or net enclosures that float on the surface of the water of lakes, rivers, ponds or coastal waters. Floatation containers used include sealed drums, sealed pvc pipe or empty plastic bottles. Styrofoam can be used, although it should be covered with more durable material to prevent deterioration. All wood used in construction should with be treated with water resilient paints or varnishes to prevent water absorption.

The dimensions of cages usually range from 1 to 2 meters on all sides. Pens are considerably larger and measure 20m x 10m. The addition of a feeding ring, which extends 10 cm down into the water, will prevent loss of food from the cage. Materials for making pens and cages will vary according to what is locally available. For example, bamboo, wooden stakes or even mesh netting may be used. It is critical that adequate spacing be provided between cages to allow waste products to be washed into the surrounding water. Cages do not have to be free floating. They can be made by using bamboo or wooden stakes constructed to form a square cage, which is firmly anchored onto the bed of the water system, or they can be made of netting to form a more flexible pen.

The advantage of cages is that they present a method for fish culture in water perhaps otherwise unsuitable for any other form of husbandry. Management of fish and harvesting is relatively simple. The disadvantages of enclosures is that fish are captive and, as a result, rely on feed from the farmer, and they are readily accessible to poachers and predators. Additionally, cages need to be monitored routinely for fouling from

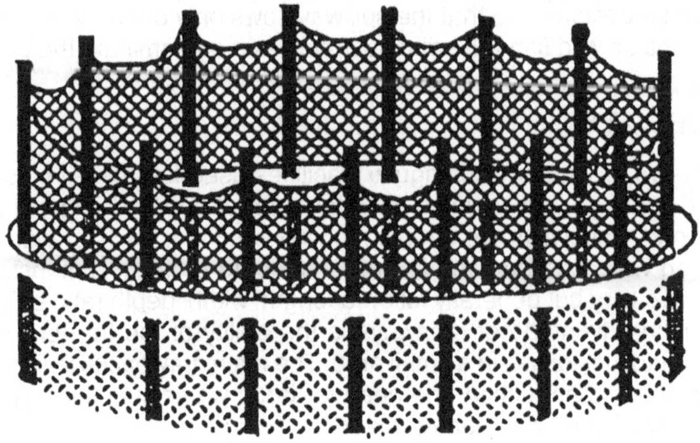

filamentous algae and invertebrate growth and any damage caused by routine use or predators. Because fish in cages are trapped, they cannot seek regions of better water quality, better temperature or avoid aggressive, territorial fish in the cage.

Net pens usually are constructed from a material that will not rot in water. Examples are nylon, plastic or treated twine. Attention to size of mesh is also important. It should be as large as possible to allow good aeration, but obviously small enough to keep in the smallest fish. Usually, pens which are staked into the water do not have a very long life but they are economical to construct and useful in areas which cannot support any other type of aquaculture.

Fish can also be reared with other integrated farming systems. For example, farmers may dig trenches or ditches in rice paddy fields. Once they have harvested the rice, they can drain the ponds and harvest the fish.

Section IV

Aquaculture Systems

There are many different ways to culture fish. The type of management undertaken depends on the type of product desired and the limiting economic resources. Operations are considered extensive or intensive. Extensive enterprises allow natural processes to regulate and provide productivity. Intensive systems are managed to produce the most fish in the least amount of space or water and are directly managed utilizing considerable human intervention and thus are labor intensive. In general, extensive operations are most appropriate small farms in developing countries. Intensive systems involve considerable monetary investment, risk and expertise. The techniques used in intensive systems vary substantially and close attention to detail is required. Therefore, intensive systems are beyond the focus of this book. Monoculture systems raise only one species. Polyculture systems raise more than one species together to more efficiently utilize all available food niches. Integrated systems combine agriculture crops with fish, animal rearing with fish, or agriculture with animal rearing with fish all in the same system. Polyculture is usually practiced in integrated systems.

To optimize management of extensive aquaculture systems the farmer must be aware of the intricacies of natural productivity and the food web. The highest amount of food energy in any system is in the **producers** (plants or phytoplankton). **Primary consumers** eat plants and constitute the highest amount of food energy among animals, and **secondary consumers** eat primary consumers and represent the second highest energy among animals. At each level of the food chain only about 1/10 of the food energy is retained. Therefore, to maximize production of animal protein, it is best to grow fish that are near the bottom of the food chain. However, there are many natural sources of food in a pond and one species can not efficiently utilize all of the food sources. Therefore, polyculture (the culture of more than one species) is more efficient than monoculture. Polyculture systems used in China are among the most efficient aquaculture systems because all fish species are low on the food chain and make use of most food sources in the pond.

Primary productivity in fish ponds can be stimulated by fertilization. In general the limiting factor in a pond is phosphorus. Nitrogen and potassium are less important components. (Inorganic fertilizers are rated for levels of nitrogen, phosphate and potassium, N-P-K. For example, 10–20–5 is 10% nitrogen, 20% phosphate and 5% potassium). Maintaining a stable pH is important for assuring availability of phosphorus. Therefore the addition of limestone to ponds with low alkalinity is necessary before fertilization is efficient. Optimum productivity for a fish pond depends on the management skills and effort that can be provided by the farmer. A good general rule for low intensity aquaculture is to maintain a phytoplankton bloom at a density that allows a visibility of 40–45 cm. This provides enough shade to prevent rooted aquatic weeds from growing and provides natural food but does not result in excess biological activity that could cause oxygen depletions under normal circumstances. In ponds with a high water flow and rapid turn over, applications of lime and fertilizers are a waste of money and productivity can only be improved by supplemental feeding.

If using **inorganic fertilizer** such as super phosphate or triple super phosphate, it is important that the fertilizer not come in contact with the mud. One good method of application is to spread the granules on a wooden platform just below the water surface. For the most rapid response, liquid fertilizer can be dispersed from a boat. Liquid fertilizer is heavier than water so it must be diluted or mixed well during application. Applications can be made at two week intervals during weather conducive to phytoplankton growth. The rate depends on the density desired but applications of 9 kg/ha of phosphate (if the fertilizer is 20% phosphate as in 5–20–5 fertilizer, 45kg would be used per application) should be used until the desired algae (phytoplankton) bloom is achieved.

In many environments **organic fertilizer** is much easier to obtain and costs less than inorganic fertilizer. It also has the advantage of supplying food energy directly to the system. Bacteria in the pond consume organic fertilizer. Zooplankton and certain fish such as common carp, tilapia or mullet then consume these bacteria. Commonly used organic fertilizers are poultry, pig, sheep and cow manure, composted vegetable matter, and non-composted vegetable matter (listed from highest to lowest percent phosphorus content, respectively). Low phosphorus fertilizers must be added to ponds at high rates and may be viewed as a low quality supplemental feeding. Caution should be used in the application of large amounts of organic matter to ponds because decomposition increases oxygen consumption and less oxygen is available for the fish.

Supplemented natural systems require the least management. They are low input systems that are utilized primarily for adding protein to the farmers diet. These systems consist of natural populations and generally involve more than one species of fish. Optimization of productivity is accomplished by elimination or reduction of undesirable species of fish, reducing predation, and partial harvesting so that a stable level of productivity is maintained. Productivity can be improved by adding natural or chemical fertilizers. Size and numbers of fish produced are inversely related. If natural reproduction results in excessive numbers of small fish, harvesting of small fish should be increased, and predatory fish should not be harvested. Any protective structures (brush or aquatic weeds) that small fish can hide in should be eliminated. Examples of this type of system include management of natural fisheries (small lakes, ponds, marshes and creeks), management of irrigation systems, management of livestock watering ponds and low intensity rice-fish agriculture. Unfertilized natural systems have productivity levels in the range of 200 kg/hectare. Fertilized systems can produce 400–500 kg/hectare surface water. A limitation in all systems involving the integration of agriculture and aquaculture is the sensitivity of fish to pesticides. By regulating the quantity per application, type, and method of application the influence on the fish population can be minimized.

Livestock watering ponds are generally of poor water quality due to small size and to suspended mud. Optimization of productivity can be accomplished by stocking the pond initially with species of fish that tolerate low oxygen and drastic temperature changes well. Additionally, strategic fencing and the use of watering troughs located away from the pond but utilizing the pond water can restrict animals from entering the pond. Prevent animals that were recently given a pesticide dip from entering the pond.

Rice-fish integrated systems make use of the natural productivity of rice paddies to produce a supplemental crop of fish. There are two systems in rice-fish culture: captural and cultural. The captural method consists of keeping whatever naturally occurring fish enter the rice field with the water when the field is flooded. Small catch basins are located in a low spot in the field, or in a group of fields. These basins are 40 m^2 by 1–2 m deep. The resulting harvest is a mixture of species and production can average up to 100 kg/ha/year. Alternatively, the type of product can be controlled by the direct stocking of desirable species of fish into the system and preventing wild fish contamination through the use of screens on the inflows. The cultural system can be concurrent, simultaneously growing fish at the same time as the rice, or rotational, where the rice and fish crops are alternated. Fingerlings or market size fish can be produced. The major species used in concurrent rice-fish culture are *Trichogaster pectoralis* (snakeskin gourami), *Clarius macrocephalus* (air-breathing catfish), *Ophicephalus striatus* (snakehead), *and Anabas testudineus* (climbing perch). The best choices are anabantids (different species of the genus *Anabas*) and clarids (different species of the genus *Clarius*) in case of low oxygen situations. The disadvantages to simultaneous rice-fish culture are the requirement for a higher water level and the limited use of chemicals on the rice crop. Particularly harmful to the fish are endrin, dieldrin and thiodan. These chemicals must not be used at all. The yield of rice is usually increased by the presence of fish because the fish eat competitive weeds and insects that cause crop damage. The fish can also reduce disease by eating mosquito larvae that cause malaria and freshwater mollusks that cause bilharzia.

Simultaneous rice fish culture is accomplished by constructing trenches or sumps in the rice fields that the fish retreat to when fields are drained for harvesting or weeding. These systems involve the use of levees and water control systems to prevent the loss of fish. The field is designed such that a trench (0.5 m wide by 0.3 m deep) surrounds the field just inside of the levee (25 cm high) and transacting trenches are added about every 0.5 ha. Also, 1 m^2 sumps are added where channels meet to collect fish during harvest. Species cultured in these systems should be very resistant to conditions of low oxygen and rapid temperature changes. Generally these fish types are the species that naturally survive in marshes and swamps. Anabantids and clarids are particularly suited for rice-fish culture in tropical climates because of their high temperature tolerance and their ability to breath air. Tilapia, Indian carp and common carp are also commonly farmed in rice fish culture.

Rotational rice-fish culture involving alternating fish and rice crops is increasingly popular in that the conditions can be optimized for each crop but rotation and diversity provides disease resistance and economic stability to the operation. This system allows optimum water levels, herbicides, pesticide and machinery use during rice production and deeper water during fish production. Structures of rotational systems are similar to concurrent systems. Bunds or levees are generally increased in height to allow deeper water for fish culture. Generally, rotational systems take advantage of the deeper water and culture more traditional aquaculture species such as tilapia and carp. Often these systems are supplementally fed and production can reach 1000 kg/ha/year.

Stocking rates for rice-fish culture

Country	***Species***	
Indonesia	*Osphronemus gorami*	stock at 5 to 8 cm; 300/ha
	Cyprinus carpio	60,000 to 100,000 1 cm fry/ha will reach 3 to 5 cm in eight weeks. 20,000 2 to 3 cm fingerlings/ha will reach 5 to 8 cm four weeks. 6000 5 to 8 cm fingerlings/ha will reach market size in 40 days.
	Oreochromis mosambicas	stock at the rate of 1000 to 10,000 1 to 3 cm fry /ha.
China	Tilapia	stock at 9 to 12 cm
Japan	*Cyprinus carpio*	60,000 to 100,000 1 cm fry/ha will reach 3 to 5 cm in eight weeks. 20,000 2 to 3 cm fingerlings/ha will reach 5 to 8 cm four weeks. 6000 5 to 8 cm fingerlings/ha will reach market size in 40 days.

Integrated animal husbandry-polyculture—The high cost of feed is the limiting factor in subsistence farming. There should be no use of food that could otherwise be used for human consumption. Extensive farming is most appropriate in these situations. Integrated animal-fish systems are the most efficient use of land and limited biological resources. However, these systems are labor intensive. The basis of these systems is livestock are fed and manure from livestock is used to fertilize/feed the ponds, or fertilize crops. In the confinement of integrated systems, pigs can be fed water plants and kitchen scraps. Water plants can be cultivated in pond margins or ditches and harvested for pig food. Bananas and rice bran can also be fed to pigs. Integrated systems can include pigs, chickens, or ducks with fish. The pig or duck manure can act as fertilizer for the crops and the fish pond. The pigs, chicken, ducks and fish are protein sources.

To maximize integrated systems, polyculture is essential. Stock common carp, tilapia hybrids, and silver carp at the ratio of 10:5:3. A minimum water temperature of 18 C or 65 F is essential to maintain productivity. Manure is used very efficiently in fish ponds, because the fish eat some parts of the manure, and what the fish don't eat acts as a food source for phytoplankton and zooplankton which are also fish food. In subsistence farming, animal manure is more efficiently used to produce a protein source (fish) instead of a plant crop. Green manure has a high biological oxygen demand that can result in low oxygen availability for the fish; it also can result in high ammonia, and possible over fertilization. Fresh manure can be added to the pond but low oxygen situations can occur, so the pond oxygen level must be monitored and low oxygen tolerant species like *Clarias batrachus* should be used. Application of greater than 100 to 200 kg/ha/day can settle to the bottom and create a high ammonia situation near the bottom muds. It is better to compost the green manure before adding it to the pond. Composting example: dig a circular pit, 1.5 meters deep, 2.5 meter diameter at the bottom, 3.0 meters diameter at the top. Layer the following every 15 cm: a silt and rice straw mixture, at a 50:1 weight ratio, pig manure, aquatic plants or green crop. The top is covered with mud to create anaerobic conditions. The pit contents should be turned over after 1 and 2 months, and should be ready for use after 2.5 months. Apply composted manure to fish ponds in small amounts so that the final rate is 5 to 10 tons per hectare per year. The majority of phosphorus is derived from feces, not urine (except for pigs which have high urine phosphorus). Phosphorus content of manure is highest from poultry (2% total dry weight), followed by pigs (1%), sheep (0.6%) and cattle (0.5%) whereas nitrogen content is 4–6% for all. The application rate of manure depends on the frequency and climate. In general, around 50kg/ha/year of phosphorus provides a good bloom. A general application rate for manure is 2.5 to 4% of the fish biomass per day.

Pig-Fish Farming

A pig produces manure at the rate of 8.2 kg/day. The manure of 20 to 30 pigs is equivalent to 1 ton of ammonium sulphate/ year. The use of 30 to 45 pigs/ha of pond will supply the organic fertilizer required for efficient fish production. In Thailand, tilapia stocked at 25,000 to 30,000 per hectare and supplemented with the manure of 60 pigs/ha, produce 2000 to 5000 kg/ha of fish in 6 months. In India, excreta from 35–50 pigs is used to fertilize one hectare of water and is applied on a daily basis. The pig manure is collected, composted and applied to the farmland and fish ponds. Pigs can be raised over the pond or near the edges.

A three tier system of poultry-pig-fish farming can be used where the poultry is raised over the pigs which are raised over the fish pond. The pigs

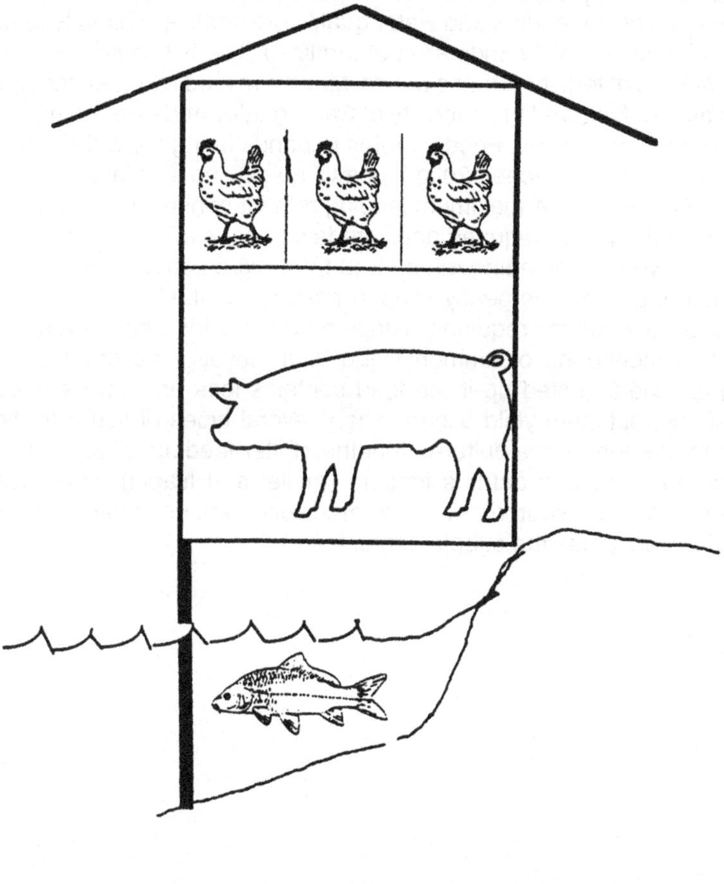

eat the chicken manure, then excess chicken waste and pig waste enter the pond acting as a fish food and a pond fertilizer (which results in fish food). A chicken produces 68 g/day of manure. The application of chicken manure at the rate of 2.5 to 4% of the fish biomass, or approximately 20 to 150 kg/ha, 6 times a week can result in average fish yields of 30 kg/ha/day.

Duck-Fish Farming

A duck produces manure at a rate of 150 grams/day. Stock ducks at a rate of 1000 to 2000 per hectare. This is equivalent to fresh duck manure applications of 150 to 225 kg/ha/day. In Taiwan, 1,000–1,500 ducks are raised/ha and produce 3500kg/ha/year of fish.

Supplemental feeding

Supplementally feeding the fish in ponds can yield high production. The intensity of feeding and water quality problems are directly related. Supplementally fed ponds are not fertilized. The fish waste serves as the source of fertility in these ponds. Maximum feeding rates depend on the type of food and the intensity of water quality and disease management being practiced. Feeding rates in ponds being fed a 32% protein ration should not exceed 20kg/ha/day if no oxygen alleviation management is practiced. A maximum feeding rate for minimal management is 40 kg/ha/day, and requires daily predawn observation of the fish during hot weather. Feeding rates of 40 to 60 kg/ha/day requires oxygen monitoring and emergency oxygen management. Over 60 kg/ha/day is intensive culture requiring nightly aeration during hot weather and routine monitoring of ammonia and nitrite levels. The rates of feeding can be adjusted up if the feed contains less protein. As in other systems, optimum yield occurs when several types of fish with different food sources are cultured together. Filter feeders (silver carp and bighead carp) and detritus feeders (mullet and tilapia) make use of the natural food sources while the omnivore such as common carp, or catfish consumes the supplemental feed.

Section V

Production Management

As previously mentioned, it is always best to start rearing fish as a source of food for a small family unit on a small scale. Once the farmer gains experience, he will be able to use his knowledge of local conditions and market supply to grow fish more intensively.

In order to be able to expand and increase the production of fish, the following questions should be asked: 1) What types of fish are available? Are they fast growing? 2) Is fish consumption already widely practiced? 3) Will raising fish be possible in conjunction with existing operations? 4) Is there enough good water available to rear fish; and can it be used cheaply? 5) Will fish be sold live or processed, and if processing is necessary, are facilities locally available? 6) Does a dependable market or use for the fish exist?

Many varieties of fish can be raised intensively. In this section, we shall deal with some of the most common species. First of all, though, we shall go through the main points of the production cycle.

Fish farms can either produce fry or fingerlings to sell to other farmers, or buy fingerlings and grow them to harvest-size fish. Of course, an experienced farmer can control both operations profitably, although working with fish at all life cycle stages usually involves several different types of ponds.

Farmers beginning fish production would be well advised to buy fingerlings. Fingerlings can also be caught or trapped. In time, it may be appropriate for farms to form a cooperative and collectively produce the fingerlings that they all need.

Care of Brood Stock

When selecting brood stock, make sure that you are in fact choosing sexually mature male and female fish that are healthy and active, and preferably fish which have proven successful in rearing healthy fish. Avoid inadvertently selecting for negative harvesting traits by spawning fish that have continually escaped harvest in a production facility. Also, do not repeatedly use a small number of spawns to generate the fish used on the facility. This could lead to inbreeding and result in genetic diseases, poor survival and poor growth. Spawning will be more successful if fish are kept in a quiet place with minimal disturbances. Broodfish should not be stocked at greater than 200 kg total weight per hectare. Brood fish should be cared for and well fed all year so the females can form and develop good quality eggs. If fish that spawn only once a year are stressed at any time during the year, they may not spawn.

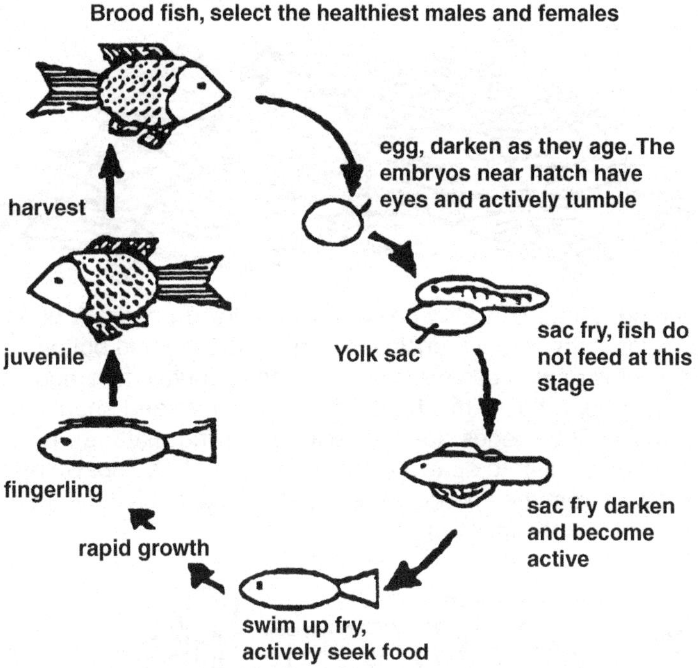

Breeding Ponds

There are two types of breeding ponds in general use:

1. Fish spawn in small ponds that have previously been drained and filled with freshwater. Artificial spawning beds or spawning mats are constructed and staked directly to or not more than 20 cm above the pond bottom. The fish spawn on the spawning mats that are then removed and placed in hatching ponds or hatching hapas. The mats are made of a dried plant material attached to a frame. The skeleton consists of wood or bamboo pieces 1 to 2 meters long and 4 cm wide with a 5 cm space between each support. Dried grasses, moss, or coconut fibers are woven into the frame. The traditional Indonesian kakaban, or spawning mat, uses Indjuk fibers. Spawning mats can also be made of bushy branches or bundles of bushy twigs. Fish that readily spawn on these mats are gourami (Anabantids), and the common carp.

Artificial Spawning Bed

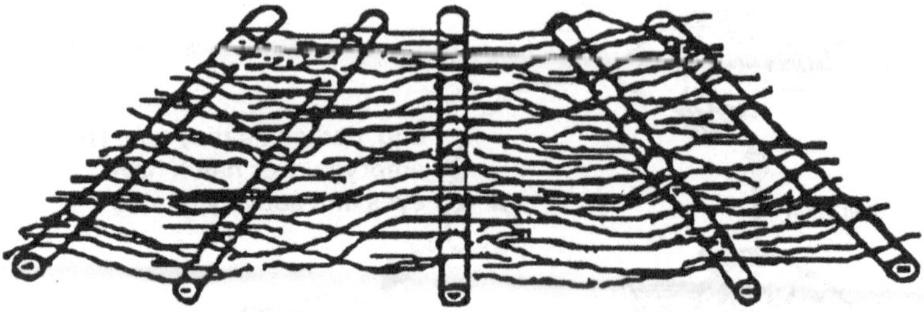

Breeding hapas can also be used in breeding ponds. A hapa is a rectangular cloth or mesh enclosure staked to the pond bottom. It can be made of varying mesh sizes and is used to contain fish, eggs or fry. A breeding hapa has a mesh size of 1.5 to 2 mm and should have a secure cover. If the eggs are adhesive, spawning mats are placed in the hapa for the fish to spawn on. After spawning, the spawning mats are removed from the breeding hapa and placed in a hatching pond or in a hatching hapa staked in a pond.

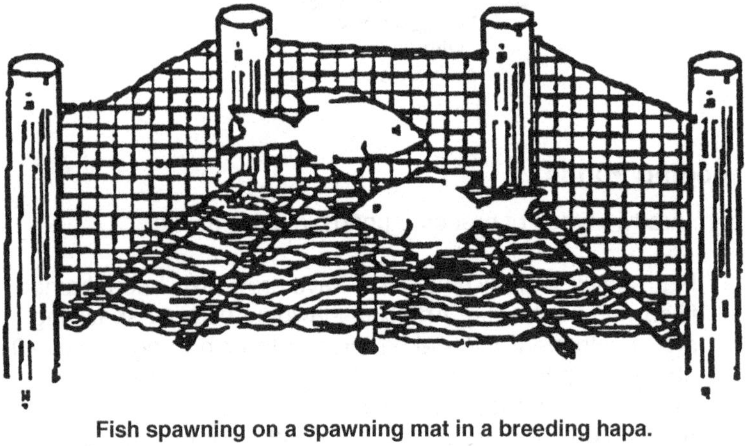

Fish spawning on a spawning mat in a breeding hapa.

2. Dubisch ponds are used in European carp culture. They are breeding and hatching ponds combined and cover several square meters. The bottom of the pond has a trough and herbage is allowed to grow on the drained pond bottom during the winter. When the natural spawning season arrives, the ponds are flooded and the brood fish are placed in the pond. When they spawn, the eggs adhere to the grass. The brood fish are then removed and the ponds become hatching ponds for the fry.

Hatching Ponds

Hatching ponds should be very small (approximately 0.05 ha) and drainable. Spawning mats with eggs adhered to them are placed directly in the pond. The eggs develop and the fry remain in the pond for several weeks. The spawning mats can also be placed in hatching hapas in the pond. Hapas used for hatching eggs typically measure 2 x 1 x 1 meters. The eggs are held in an inner chamber that has a mesh size 2 to 2.5 mm. When the fry hatch, they fall through to the outer chamber and the egg debris is retained in the inner chamber and is removed after all the eggs have hatched. The mesh size of the outer chamber must be small enough to retain the fry and is species dependant, but generally 1.0 mm is sufficient. Yolk sac fry get their food from their yolk sac. As it is absorbed, the fish become more active and change from being transparent to a darker color. At this stage they start to look for food and it is important that they learn to feed very quickly to ensure continued good growth. If a hatching hapa is used for any type of fish, supplemental feed should be given at 3 days after hatch, and it is critical that the fry be released 4 days after hatch. At three days post hatch, the fry still have some yolk reserves and begin searching for food. At 4 days post hatch, they must have natural food. There may be a large mortality at this stage (60–80%), both from fry unable to find food and from predators (other fish, birds, aquatic organisms). Hatching ponds should be well dried out before using them to make sure they are free of undesirable fish before the spawning mats are placed in them. The ponds are then fertilized, preferably with organic fertilizers such as manure or plant matter to promote the growth of plankton and zooplankton that are natural highly nutritious food essential for rapid growth of the fry. A natural food web will take 1 to 2 weeks to become established in the pond. When the eggs are laid, the pond should be filled with water and organic matter added. This will ensure that there will be natural food available for the fry when they first require a natural food source. The young fry can be supplementally fed with small amounts of rice bran or other locally available food. It is important to feed these fish small amounts often. If you plan to transport fry, you will need buckets and plankton net or fine mesh sieve cloth on hand to make nets and hapas.

Most eggs can be transported up to one day if they are kept wet or in water with oxygen and handled very carefully. In tropical areas, the transport water must not be allowed to heat up. Larvae and sac fry can be transported at the rate of 1000 per liter of water with oxygen and very gentle handling. Feeding fry can be transported at the rate of 1000 per liter of water with oxygen.

Fingerling Ponds

After a few weeks, fry can be transferred to fingerling ponds approximately 0.05 ha in size. They may also remain in the hatching pond until they are large enough to be placed in a rearing pond. Once again at the fingerling stage mortality rates are likely to be high. Do not crowd the fish at this stage—give them plenty of fresh, well oxygenated water and, if giving supplemental feed, then feed small amounts often. Fertilizers can be added to the pond to increase natural food available for the fish. Fry are typically stocked into a fingerling pond at the rate of 100,000 to 200,000 per hectare. The transportation rate for fingerlings up to one month old is 100 fingerlings per liter of water, supplied with oxygen.

Growing Ponds

Finally, when the fish are well established and appear to be growing well, they are placed in the growing system (raceway, pond net pen, cage, etc.). If possible, ponds should be dried out and filled with fresh water to ensure there are no predacious fish prior to stocking. Stocking densities of fish vary enormously; typical rates are 10,000 to 30,000 fingerlings per hectare if supplemental food is supplied. With no feeding and low level fertilizing, stocking fish should be limited to yield a harvest weight of 70–110 kg/hectare. If the fish will average 450 grams at harvest, 150–250 fish should be stocked per hectare. Do not exceed 1000 fingerlings per hectare if no supplemental feed is supplied. Well fertilized ponds can yield 1000 kg/ha/ year and fed ponds yield 2000 kg/ha/year at low to moderate feeding rates and stocking rates are adjusted accordingly.

Selected Species

Tilapia *(Oreochromis* species)

Tilapia are very fast growing fish and in many countries in Africa they are harvested after only six months. It is considered an ideal fish to provide a cash crop. They mature very rapidly. The fish can reach sexual maturity in six months, and once mature, can reproduce almost continuously. There is a potentially high risk of hybridization between the different species and this may produce stunting problems. A management practice adopted by some farmers is to introduce a few predacious fish, such as walking catfish, to reduce the production of small fish. Most Tilapia are "mouth brooders". After fertilization of the eggs, the female takes the eggs into her mouth, where they develop. A 100 to

200 gram female tilapia can produce 200 to 400 eggs at each spawning. The females form eggs every 30 days. When the fry are free swimming, they use their mother's mouth as a retreat from predators. Tilapia produce fewer eggs than other kinds of fish but this protective mechanism ensures a higher survival rate. For spawning, a pond with a loose bottom is generally preferred. The brood fish should be stocked at a rate 1 male: 3 to 4 females per square meter. If it is not possible for the male Tilapia to make a hole (the nest) in the pond bottom, large jars, crates and boxes can be separated in the pond as artificial nests. Fingerlings can be stocked at rates of 5,000 to 20,000/ha.

The problems outlined above related to rapid reproduction can be prevented by placing predacious fish in the ponds to eat the young fry, allowing the adults to keep on growing (the predacious fish should not be so big that they can also eat the adults) or by producing monosex (all the same sex) fish. This is accomplished by feeding a blended hormone diet feed (17 methyltestosterone 3 mg/kg food) to fry from the first stages of feeding at 12–15% body weight per day. The hormone treatment is probably only possible when organized by government or other organizations but it is worthwhile if these facilities are available and if this process is approved for food fish in your country. In Jamaica, feeding supplemental hormones produced 99% males in ***Oreochromis niloticus*** after 25–28 days, and the same feeding regime produced 95% red Tilapia males after 30 days. Culturing Tilapia in floating cages has also been reported to substantially reduce reproduction because there is no substrate available for spawning.

Tilapia can be difficult to sex. The male has a single small opening at the tip of the genital papilla. The female has two openings at the tip of the genital papilla. One is a small ureter and the other is a slit like oviduct. Distinguishing males from females can be facilitated if the urogenital area is swabbed with an inert coloring agent such as food coloring.

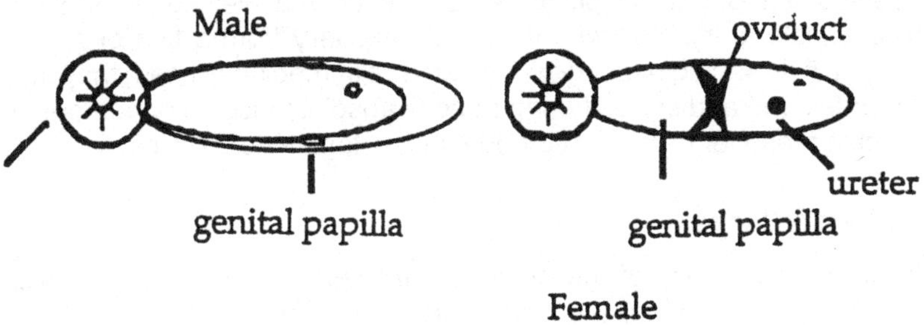

Tilapia urogenital areas

Carp

The main types of carp are Chinese, common, and Indian. All the carps grow well in polyculture systems and in integrated animal-fish systems. Common carp naturally spawn when water temperatures are 18 to 22 C and the terrain is inundated. The Chinese and Indian carps naturally spawn in response to the flooding of the rivers during the monsoon seasons. In ponds, holding the brood fish in shallow water and then rapidly raising the water level of the pond can simulate this. When buying new stock for a facility, use fish that have no history of spring viremia of carp, carp pox, or grass carp reovirus. These are devastating diseases in carp culture.

The common carp (*Cyprinus carpio*) is an omnivore. Therefore it utilizes plants, insects, small worms and detritus as food. They are sexually mature at 1 to 3 years of age and will reproduce naturally in ponds. Common carp will spawn under most conditions. Spawning is generally allowed to occur naturally in early spring (water temp 20°C) in ponds stocked with fish weighing 1 to 2 kilograms. Spawning can often be triggered in a pond if the water level is increased to simulate spring floods. These fish can be spawned in Dubisch ponds, or in breeding ponds using spawning mats, then transferring the eggs to a hatching pond. Common carp eggs are adhesive and one female can produce 50 to 60,000 eggs per kg of body weight which will result in 8000 to 15,000 post larvae. Egg incubation time is 3.5 to 4 days and just hatched larva measure 4.8 to 5mm.

Spawning can be aided by injections of carp pituitary collected from other fish. The pituitary gland is also called the hypophysis, so this procedure is referred to as hypophysation. This extract can be prepared by removing the pituitary gland from another fish and homogenizing it in a small amount of water (measured in drops). The preparation should be filtered in some way, and then injected into the fish. The gonadotropins in the gland are water soluble and act to stimulate spawning in the injected fish. Use whole pituitaries ground up and injected into the fish. The female is injected with the whole pituitary from a fish of similar weight. If dissected pituitary glands are not immediately used, they may be preserved and stored for later use in absolute alcohol or acetone. A stocking ratio for most carp brood fish is 1 female to 2 males.

Chinese Carps

This group consists of the bighead carp (*Aristichthys nobilis*), which feeds on zooplankton, the silver carp (*Hypophthalmichthys molitrix*), which feeds on phytoplankton, the grass carp (*Ctenopharyngodon idella*), which feeds on aquatic vegetation, and the mud carp (*Cirrhina molitorella*), a bottom feeder which feeds on detritus. These carp exist well in polyculture systems because each occupies a different food niche so maximum utilization of the aquaculture system can be used. Yields of 7500 kg/ha/year can be obtained with fertilization of Chinese Carp polyculture systems. Chinese carp can be intolerant of handling (grading, netting, etc.), so great care must be taken while they are being handled and they may need to be anesthetized.

Chinese carp will spawn once a year under normal conditions. The broodfish can be held together, but must not be overcrowded. The maximum total weight is 183.6 kg/ha (1000 lbs/ac); a good combination is shown below:

grass carp	280 kg/ha or 250 lbs/ac
silver carp	561 kg/ha or 500 lb/ac
bighead carp	280 kg/ha or 250 lbs/ac

All Chinese carp eggs are non-adhesive, slightly demersal, and will float following the riverine current. At hatch, the larvae are 4 mm long and move with the water flow. Eggs and larvae are collected from the wild, then moved to rearing ponds. However, the most common method of propagation for this group is manually collecting the eggs and milt (stripping) and then incubating the fertilized eggs. If the eggs are incubated and hatched in a tank or trough, make sure to place fine screens over water exits and monitor the screens frequently for cleaning. For stripping, suitable hormone combinations in proper amount and correct sequence to mature ova or induce ovulation, must be administered. These specifics are included in the information for each species. Proper recognition of ovulation is very difficult for the inexperienced person. A good method is to put males into the holding tank with the females and watch their pre-spawning behavior and attention to the females. After the female ovulates, the eggs can be expressed from her abdomen by applying gentle pressure. They are collected in a dry bowl directly from the fish. Milt, or sperm from the male is collected in the same manner directly onto the eggs in the bowl. The eggs and sperm are gently mixed for a couple of minutes. Water is then added and this mixture is gently mixed for two minutes. The eggs can then be transferred to hatching jars or hatching troughs for incubation.

Chinese carp fry initially feed exclusively on zooplankton, so pond preparation is very important (see section for fry pond preparation). Stocking rates for primary rearing (first feeding to 5 cm) of Chinese Carp fry are:

grass carp	1 to 1.25 million per hectare
bighead carp	1.25 to 2.5 million per hectare
silver	1.5 to 2 million per hectare

The bighead carp spawns at water temperatures of 18 to 23 C in temperate areas and 25 to 30 C in the tropics. These fish reach sexual maturity at 6 to 8 years of age in temperate regions, and 3 to 4 years of age in tropical regions. Induced spawning of the females requires either 2 to 3 milligrams of dried or fresh cyprinid pituitary per kilogram of body weight, or 700 to 1000 IU human chorionic gonadotropin per kilogram of body weight. Their eggs will hatch in 1 to 1.5 days at 23 to 27 C. Just hatched larva measure 5 to 5.2mm. The bighead carp naturally feeds on zooplankton, phytoplankton, rotifers and detritus and will also eat rice bran, peanut cake and soybean cake. Fertilizing ponds will increase the natural food supply and subsequent weight gain. The usual growth obtained is 150 g the first year and 900 g the second year.

The mud carp eats plankton and minute particles of detritus. These fish reach sexual maturity at 3 to 4 years of age and 1 kilogram of weight. They spawn at 26 to 30 C and respond to hypophysation (carp pituitary). The females can be stripped 16 to 18 hours after administration of carp pituitary extract. Reported growth rates are 75 g in the first year, to 300 and 600 g in the second and third year, respectively.

The grass carp, or white amur, eats aquatic vegetation. They reach sexual maturity at 3 to 5 years of age in tropical regions and 4 to 7 years in temperate regions. Grass carp spawn at 22 to 25 C and respond to intraperitoneal or intramuscular injections of 3 to 4 mg of carp pituitary per kg of female body weight. Males and females are used in a ratio of 1:1 and can be spawned in a pond, hapa, or stripped. Females will produce 60,000 to 80,000 eggs per kg of body weight. Egg incubation time is 24 to 32 hours at 22 to 25C. Just hatched larva measure 5 to 5.2mm. Fry require zooplankton and phytoplankton and must be transferred to properly prepared fry ponds. They can also be supplementally fed. Grass carp are typically stocked at low rates to control aquatic vegetation, but can be a significant part of the polyculture yield.

The silver carp eats zooplankton as fry and fingerlings and converts to phytoplankton as adults. They filter particles 17 to 300 microns in size and will eat supplemental feed of rice bran, peanut cake, and soybean cake. They reach sexual maturity at 2 to 3 years or 2 to 4 kilograms in tropical regions and 4 to 6 years in temperate regions. They spawn at 21–26 C and respond to hypophysation, human chorionic gonadotropin (HCG), and lutenizing hormone releasing hormone (LHRH) and can be stripped. The silver carp is the most sensitive of these species to handling stress. A female can produce 50,000 to 60,000 eggs/ kg of body weight. Incubation time is 24 to 32 hours. Just hatched larva are 5 to 5.2mm in length and require zooplankton at first feeding.

Indian Carps

The Indian major carps are the Catla (*Catla catla*) which is a surface and water column feeder and eats zooplankton, the Rohu (*Labeo rohita*) which is a browser, and the Mrigal (*Cirrhinus mrigala*) which feeds on bottom detritus. These fish are frequently stocked in polyculture systems. Indian carps are not as hardy as Chinese carps but can still be produced successfully. These carps are semi-tropical and require indoor over wintering in temperate climates. Indian carps spawn after monsoon flooding (June through August in India), and can be induced to spawn using hypophysation. These fish can be spawned in a Bundh, a special breeding pond where riverine conditions are simulated, a breeding hapa, or stripped. The eggs are semi-floating and adhesive. The fry require zooplankton, so the fry pond must be properly prepared as previously described. All types of fry are typically stocked at a rate of 1 to 2 million /ha.

The catla reaches sexual maturity in 2 years and spawns at 30 to 32 C. They respond to hypophysation at the rate of 3 to 4 mg/kg body weight with the female receiving two doses, and the male receiving one dose at the time of the female's second dose. The female will produce approximately 100,000 to 250,000 eggs per kg of body weight. Egg incubation time is 15 to 20 hours at 27 to 31C. Just hatched larva measure 4.4 to 5.3mm. Fry may be supplementally fed with ground nut oil cake and rice bran. The catla will grow up to 1.5 and 5 kilograms in the first and second years, respectively.

The rohu reaches sexual maturity at two years of age in India, but requires three years in Bangladesh. They respond to hypophysation. The number of eggs produced varies from 200,000 to 400,000/kg of body weight. Incubation of eggs requires 14 to 20 hours, and just hatched larva are 3 to 4.5 mm in length. Fry can be supplementally fed as for the Catla. Growth rates under culture conditions are 680g to 900g at the end of the first year and 2.5 to 5kg at the end of the second year.

Mrigal males and females reach sexual maturity at one and two years of age, respectively. Females produce 150,000 to 300,000/kg of body weight. Egg incubation is 16 to 24 hours at 24 to 31C. Just hatched fry measure 3.5 to 4.8mm and should be supplementally fed as for the catla. Under culture conditions, the mrigal can grow to 1 kg at the end of the first year and 2.5 kg at the end of the second year.

Milkfish (*Chanos chanos*)

Two hundred and fifty thousand (250,000) tons of Milkfish are estimated to be produced yearly in the Philippines, Indonesia, and Taiwan.

The major constraint is that the fish cannot be bred in captivity. Milkfish mature at 4 to 6 years of age, or 1 meter in length and spawn in marine waters at temperatures of 23 to 32 C and depths of 25 meters. The fry are collected in shallow coastal waters in April, May and June and distributed to farms through out the coastal regions. The fry are reared in nursery ponds until they reach a total length of 30 to 60 mm. The nursery ponds measure 1/2 hectare to several hectares with depths of 0.3 to 1 meter and are usually constructed so that one end of the pond can drain directly into coastal saline waters. After two months, the fingerlings are moved into grow-out ponds. In Indonesia, young fry are initially stocked in nursery ponds at 500/hectare and reach 200–400 g after two to four months in a grow-out pond. In Taiwan and Malaysia, polyculture systems of milkfish and the large prawn *Panaeus monodon* have been very successful.

Giant Gourami (*Osphronemus goramy)*

This fish is a member of the family *Anabantidae*, also called Anabantids. These fish are air breathers and are suitable where low oxygen situations are prone to occur. They are cultured in China, Ceylon, India and the Philippine Islands. They usually require 2 to 3 years to reach harvestable size but bring a high market price. Giant gourami reach a maximum length of 65 cm and are sexually mature at 2 to 3 years of age when they weigh 2 to 3 kgs. Brood fish are conditioned on natural food, rice bran and minced soft plants. The giant gourami builds an egg nest. Artificial nests can be built and placed in the spawning ponds, then later removed to collect the eggs. The best spawning ponds are 300 to 600 m^2 and 1 m deep. Gravid adults are selected on the basis of full round bodies and reddened fins. They are stocked at the rate of 2 females and 1 male per 7 to 10 m^2, with one artificial nest per female. The artificial nests are made of 75 to 90 cm long stick bundles tied together at one end and open at the other end. It is shaped like a funnel and is 30 to 35 cm wide at the end. Bunches of fibers are fastened or stuffed inside (the same type of fibers that are used to make spawning mats). It is placed in a horizontal position, 20 cm from the surface. The nests should be inspected for eggs every 2 days. After the fish have spawned, there is surface oil on the water, and a characteristic fishy smell. It typically requires about 2 weeks for the fish to spawn after being placed in the pond. After spawning, the nests are collected and placed in a hatching pond, or the eggs can be gently removed and placed in a hatching pond. There are 3000 to 4000 eggs per nest, and they hatch in 2 to 3 days in tropical water temperatures. The feeding fry require zooplankton, so hatching pond preparation should begin as previously discussed at the same time as the nests or eggs are placed into the breeding pond. At two weeks of age, supplemental feeding of the fry can begin using 1 cup of white ants or 2 cups of peanut waste per 3500 fry per day. Fry are then transferred to rearing ponds at rates of 1000 to 1500 fry/ha and fed minced floating plant leaves. A size of 5 to 8 cm should be attained in 5 months. At this time they are stocked into stagnant ponds or integrated rice-fish systems at the rate of 300 fish/ha.

Stocking Densities And Growth Potential

Unfortunately, losses will always occur when rearing fish. However, when stocking fish, the following management procedures should be adopted to minimize mortality.

1. Maintain good oxygenation.
2. Avoid handling the fish too much.
3. Never move, grade, or handle fish when water temperatures are very high. It is always best to handle fish early in the morning, not in the afternoon.
4. Always have equivalent water quality parameters when moving fish from one holding system to another. A temperature difference of more than 2° C (what you can detect by touch) must be tempered before stocking. This involves slowly changing the temperature of the water that the fish are in by adding water from the pond that they are being stocked into. The rate of temperature change should not exceed 2° C in 15 minutes.

If fish are stocked too densely, growth will be inhibited. If under stocked, then there is under utilization of the environment and this reduces profit for the farmer. Stocking rates will vary according to the type of fish grown, the environmental conditions and the kind of system the fish are held in. Below are specific examples of recorded stocking densities in various countries.

1. JAVA, MADURA
   Ponds: average yields in minimal manage polyculture systems per hectare were: Chanos, 150 kg; Prawns, 50 kg; and various carp species, 25 kg.
2. INDONESIA
   Ponds: Milkfish, fingerlings stocked at 500/hectare may reach 200–400 g in eight months.
3. INDIA
   Ponds: Indian carps were stocked at 4,000–11,000 fingerlings per hectare. The ratio of *Catla* to *Rohu* to *Mrigal* was 3:4:3.

4. CHINA
   Ponds: Common carp and Chinese carp were stocked at 10,000–20,000 fingerlings per hectare.
5. CENTRAL AFRICAN REPUBLIC
   Tilapia nilotica and *Clarias lazera* initially stocked 20,000/hectare and 1,500–10,000 hectare, respectively. In 0.34 hectare ponds, production rates vary from 6,500–11,484 kg/hectare/year with and without supplemental feeding. In six months, the mean weight of Tilapia was over 100 g; the mean weight of *Clarias* was over 200–300g.

A general guide to stocking of fingerlings in ponds is:

Common carp	1,000–10,000/hectare
Tilapia	1,000–20,000/hectare
Indian carp	1,000–10,000/hectare

FEEDS AND NUTRITION

Adequate nutrition is one of the most important concerns in the production of healthy fish in supplementally fed ponds. Evaluation of food for its economic benefit is based on calculating a food conversion ratio. This is the amount of feed (in kilograms) required to produce one kilogram of fish. A feed conversion ratio of 1:1 means that for each kilogram of food fed, a kilogram of fish was produced. High quality food gives a lower ratio than poor quality food. The cost of feeding fish must be offset against the price expected for fish at harvest. Feeding may account for 50 percent of operating costs. It does not need to be so expensive if a lower feed conversion ratio is acceptable using locally available materials. While some fish require a high protein feed, others are able to thrive on the natural food in the pond.

Nutritional deficiencies—are most often characterized by poor growth, poor feed conversion and a high incidence of infectious diseases. The most common nutritional deficiency seen in intensively fed fish systems is vitamin C (ascorbic acid) deficiency. Severe cases are characterized by broken back syndrome (spinal deformities) deformed gill filaments or broken gill isthmus. Causes of vitamin C deficiency are improper formulation, and improper storage. Vitamins C and E break down rapidly especially in hot humid conditions. Other deficiencies lead to anemia, fatty livers and unusual pigmentation. Deficiencies are rare in ponds that rely heavily on natural foods but in ponds that are heavily dependent on supplemental feed, formulations designed for the fish being cultured should be obtained from a reputable manufacturer.

1. Formulated Feeds

Feeds are formulated by blending several different nutrients, making it possible to prepare a nutritionally adequate feed at a reasonable cost. Below are a few of the common feeds that are often blended to prepare fish rations.

Many of these feeds may not be available locally. These examples will help to identify local sources of equally good nutrition.

Fish meal: First rolling and pressing cooked fish to remove water and oil, and then drying it produce fish meal. Such species as menhaden, anchovy, or shad make excellent fish meals that are high in essential amino acids. Other locally available fish may also be used as a substitute for this. These meals are highly palatable and can act as an attractant when mixed with other ingredients. Fish meal can provide up to 8 percent of the ration.

Meat and bone meal: This is the dry rendered product derived from mammalian tissue, excluding hair, hoof, horn, manure, and stomach contents. It is a nutritious protein source, containing essential amino acids that can be incorporated with plant proteins. Because of its variable quality and high ash content, meat and bone meal should be limited to 15 percent of the diet. It may also be used in combination with, or in place of, fish meal.

Fish offal meal: Fish offal meal is produced by cooking offal (waste products), pressing it to remove water and oil, and then drying it. It may contain up to 50% protein, and is a good source of calcium and phosphorus. Offal meal may vary in quality depending upon the raw products used to prepare the meal.

Soybean meal: This is a major protein of commercial catfish feeds in the United States. It is high in essential amino acids and is both digestible and palatable. Soybeans must be chemically treated by solvent extraction and roasted to inactivate anti nutritional factors.

Peanut meal: This has been used to replace a small part of the soybean in feeds. It has a low lysine content (an essential amino acid). Also, peanuts are susceptible to contamination with the mold ***Aspergillus flavus***, which produces **aflatoxins**. Never use moldy peanuts or other moldy foods.

Cottonseed meal: This is used as a part of a ration in locally produced fish feed. It may contain **gossypol**, a toxin that can depress growth in fingerling fish.

Maize: Known as corn in the United States, this can be used as a major energy source in fish feeds. Cooking during the pelleting process will improve its digestibility. Corn can form up to 40 percent of the total diet.

Wheat: This is very palatable and digestible and has a high nutritional value similar to maize. It is generally an expensive addition to a diet, however.

Wheat middlings: This is a by-product of the milling process. The middlings have a high amino acid content but are usually more expensive than other milling products.

Rice bran: Rice bran is composed of rice milling by-products. It is a good energy source but is low in most essential amino acids.

Other locally available feeds should be investigated for their potential use in rearing healthy fish. For example, in the Central African Republic, *Tilapia nilotica* were grown in concrete tanks using locally available materials. The following feed conversion ratios were obtained: groundnut cake 3.6:1; cottonseed cake, 4.8:1; cottonseed 18.9:1; and brewery wastes, 12.6:1. It can be seen that conversion ratios are lower when the protein concentration is higher. In some parts of Africa, tilapia can be grown on organic fertilizers and rice bran (or cottonseed meal—depending on availability). Yields can reach four to six kg/hectare in six months.

Feeding Types And Strategies

Floating Versus Sinking Feeds

Fish feeds are made in floating and sinking forms. The type used is based upon the farmer's preference and the feeding pattern of the fish. The feeds are of similar composition.

The main advantage of floating feeds is that the farmer can see the fish feeding and can monitor their health more easily. Sinking feeds generally cost less than floating feeds. Strong winds will not blow the feed across the pond. Some species of fish, such as bottom-dwelling catfish, prefer sinking feeds.

Particle size: The size of the pellet depends upon the size of the fish's mouth, and mouth size varies with age. Generally, the larger the fish, the larger the pellet. Where there is uneven fish growth, or different age groups, a mixture of two or more sizes of feed may be necessary. In mixed size populations, a swim-through feeder may be required to allow smaller fish the opportunity to feed. An underwater cage is placed in a pond with feed inside it. Small fish can pass through the mesh to reach the feed, but larger fish are kept out.

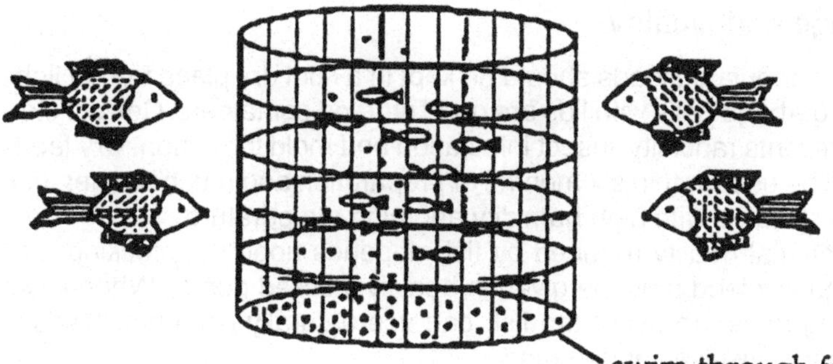

Homemade Feeds

Homemade feeds can satisfy the nutritional needs of farmed fish. Locally available feeds must be fully investigated to determine their potential nutritional value in the diet. Two simple feeds, which take advantage of inexpensive animal protein, have been developed at Southern Illinois University. These formulations are included to demonstrate how locally available products may be adapted to produce an economical fish food. Where pelleting is impossible, dough of corn flour, peanut meal, and soybean meal can be made locally. In one preparation, whole soybeans were roasted for 15 minutes at 218 C and ground by being passed through a food chopper using a 3mm plate. The soybeans were then passed through a chopper (1 hp chopper, Toledo Scale Company) with whole frozen but cooked gizzard shad (this can be substituted with another readily available forage fish), one percent vitamin mix, and 2 percent salt mixture. The mixture was ground three times to ensure complete mixing. The chopper extruded a 3 mm diameter strand of moistened feed that was broken into pellet form after drying.

Feed	*Ingredients*	*Percent of Total (Wet Weight)*	*Percent of Total (Dry Weight)*
Feed 1	Ground roast soybeans	48.5	79.5
	Fresh/frozen gizzard shad	48.5	16.0
	Vitamin mix*	1.0	1.5
	Salt mixture USP XIV	2.0	3.0
	TOTAL	**100.0**	**100.0**
Feed 2	Ground roasted soybeans	32.3	52.5
	Corn meal	16.2	27.0
	Fresh/frozen gizzard shad	48.5	16.0
	Vitamin mix*	1.0	1.5
	Salt mixture USP XIV	2.0	3.0
	TOTAL	**100.0**	**100.0**

*Obtained from Nutritional Biochemicals, Cleveland, Ohio.

Storage and Quality

Ideally, formulated feeds should be kept in a cool dry place free of light. Clean garbage pails with lids are good storage containers. Correct storage prevents rancidity, insect infestation and mold formation. Dry feeds should be used within six months of preparation and much sooner than that in climates with high humidity and high temperature.

Nutritional quality required by fish depends upon the stocking rate. High quality feed must be given in heavily stocked ponds. When lower stocking rates are used, the fish can use naturally occurring feeds in the pond as well as added feed.

Fry feeds: In most aquaculture systems fry eat natural sources of food. Generally, supplementally fed fry must be fed a complete diet of 40–50% protein, or supplemental feeds plus an added source of nutrition (such as ground liver). Half of the protein should be supplied by animal protein. From the stage of first feeding, after the yolk sac has been absorbed, dry particles of food should be offered. If the fry are in troughs or tanks, they should be fed frequently (every few hours). There will be some wastage of feed. This should not be allowed to accumulate. The growth phase of fry is very rapid and it is important to get as much growth at this stage as possible.

Fingerlings: This is similar to fry feeding although larger size pellets should be offered. Protein content of 26–38 percent is adequate.

Production fish: Larger fish are fed according to appetite levels, water temperature, stocking density, and water quality. As the fish get bigger, the pellet size offered should increase. Food should be spread over a wide area to allow all fish the opportunity to feed.

Broodfish: These fish should be fed 4 to 5% of their body weight per day. They must be in a state of good health and steady weight gain to maintain high egg quality in the females. They should also be stocked at a lower density than production fish.

Feeding Rates

Only give fish food which can be consumed in 10–15 minutes at each feeding time to minimize waste and maximize consumption. Fry and fingerlings less than 3 cm in length should be fed 15% of their body weight, fingerlings 3 to 15 cm in length should be fed 10% of their body weight, and fish greater then 6 inches, 3% of their body weight. To determine how much total food to feed a tank or a pond, perform this calculation:

total number of fish x the average weight per fish x percent body weight to be fed (3, 10, or 15%) = total amount

Feeding rates should be reduced in cold water (fish metabolism is slower at lower temperatures), when fish are sick, when fish have an abundant natural food supply, or when there has been a stress in the pond. Causes of stress include partial harvesting and periods of low oxygen on hot, still, or cloudy days. Alternatively, feeding rates can be increased when natural food is not available or when fish are held in low numbers and with increasing growth of fish. In ponds with high stocking densities, maximum feeding rates are determined by the water quality. Farms with low levels of management and oxygen monitoring equipment should not exceed a rate that causes the phytoplankton to become denser than a visibility of 40–45 cm.

Time and Frequency of Feeding

Never overfeed fish. Feed that is not eaten may act as a pollutant and lead to water quality problems. Small farms may choose not to calculate feed rates but to watch how fish feed and provide just enough for them not to leave any waste. Feeding should not be carried out in periods of low oxygen. In many fish production areas, feeding begins at midmorning to allow a long enough period of sunlight so that photosynthesis can provide adequate oxygen in the pond for feeding and digestion. Some farmers prefer to feed smaller amounts twice daily.

Always try to feed from the same location on the pond bank. This will train the fish to come to the same area for feeding. The feeding site should be on the windward side so that feed will float over the pond and not directly up to the pond edge. Similarly, if there is a water flow (a raceway, for example), feed should be given at the upper end so that feed is gradually washed down over the rest of the system. Food should be thrown into areas that are 1–1.5 m deep.

Harvesting

Harvesting should always be done when there is plenty of aeration and when water temperatures are not high, usually early in the morning. It is advisable not to feed for one or two days before harvesting. Methods of harvesting generally include some draining of a pond and letting fish gather at the end of the pond (especially if there is a deep end). The pond or tank can be seined with a net that is weighted at the bottom to prevent it from rolling up and allowing fish to escape.

Management Guidelines

Routine maintenance checks on a farm will help to avoid major problems.

Daily Management

1. Check all inflows and outflows to the system.
2. Check for changes in water color (indicating possible algae die-off conditions).
3. Precautions should be taken to guard against predators.
4. Determine if fish are feeding well, behaving normally and look for dead or diseased fish.

The farmer should inspect the ponds daily and do this when oxygen levels are lowest (at sunrise) to monitor the oxygen availability for the fish, and again when the fish are fed.

Occasional Management (every few weeks)

1. Check walls and dikes.
2. Cut grass on pond banks—This can be fed to grass carp or used in compost heaps. Keeping the banks clear helps eliminate hiding places for predators.
3. Check the fertility and turbidity of the pond. If necessary, fertilizer (organic and/or inorganic) should be added.

Section VI

Fish Health Management

Fish health management involves the careful balance of healthy fish, a good environment, and pathogens that may harm the fish. All of these things must be managed smoothly and efficiently to avoid disease, or to reduce their severity.

ENVIRONMENT

Water Quality

Optimum water quality should always be maintained. Low dissolved oxygen, high ammonia, and high nitrite are all stressful to fish. Careful monitoring of these parameters can alleviate poor water quality stress. Low oxygen, for example, can be corrected by aerating the water. Addition of salt (sodium chloride) to the water can alleviate nitrite toxicity.

Some water quality problems cannot be prevented. However, constant evaluation of the water may signal a developing problem and corrective measures may be taken at an early stage. We have already discussed turbidity and color as a sign of a healthy pond. In some areas, it may be possible to also test the other parameters.

Reservoirs for diseases

Wild fish may get into fish ponds and introduce disease. They also will compete for food with the stocked fish. Screening inflowing waters should help prevent predacious fish from entering the system. Predatory birds are also reservoirs for some important parasites in fish and can transmit diseases from pond to pond. Continuous culture systems involving partial harvesting and restocking on a continuous basis can lead to severe disease management problems because older fish serve as a reservoir for disease causing agents that can infect the younger fish.

Before stocking a pond, the bottom should be dried thoroughly and then the pond refilled with fish-free water. This eliminates wild fish and fish that escaped when the pond was last harvested. These fish are a source for disease organisms and allow pathogen carry-over in the pond between crops of fish. Frequently during this resting period vegetation can grow in the pond and provide a natural food source for the fish or the invertebrate food organisms. Also this is a good time to add agricultural lime if needed.

If it is impossible to drain the pond before stocking with fish, those fish already in the pond can be killed with various agents, such as rotenone. In India, ponds can be cleared of unwanted fish using an oil seed cake locally called Mahua. The oil cake contains Mowrn, an alkaloid that is highly toxic to fish, mollusks, and other aquatic animals. The advantage of this method is that the toxin undergoes degradation within about

10 days. Another method from India uses ***Bassa latifolia*** cake applied at 250 ppm. Apparently this treatment not only kills unwanted fish but also acts as an organic fertilizer.

Finally, if other chemical agents are available, they also may be used if all instructions for their administration are followed, examples include rotenone and chlorine. It is important to stress that great care should be taken when handling these products both in terms of human safety and for the consequences of these agents draining through irrigation ditches. They should not be used in high porous soils where water leaks out through dikes.

Stocking density

A high rate of stocking of one species of fish optimizes the chances of an infectious disease being transmitted between fish in the pond. Therefore, a heavily stocked pond of one species of fish has a much greater chance of having a devastating disease outbreak than a pond stocked at a moderate rate with multiple species of fish (polyculture).

Disease

For a farmer to effectively manage his fish he must know the initial signs of disease, emergency measures that can be taken and finally how to get help. It is very difficult to tell when fish are not feeling well. As

Behavior indicative of disease

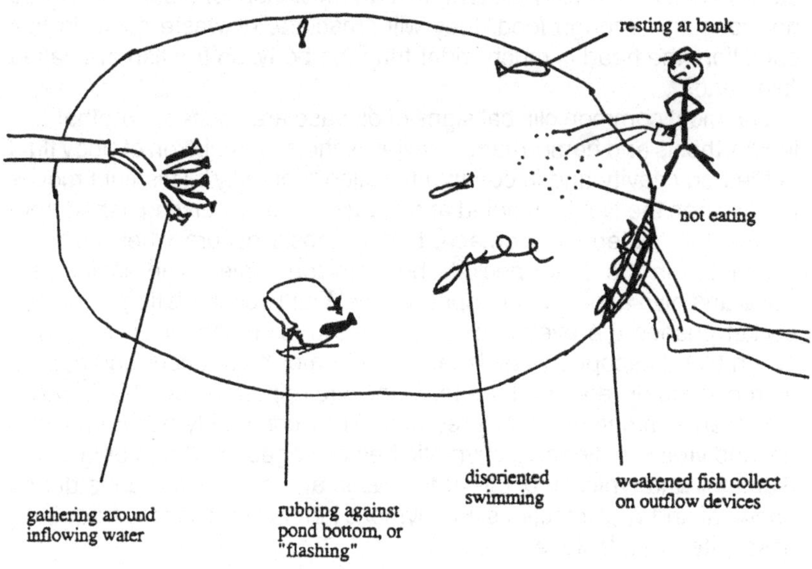

with land animals, sick fish are generally less active. However, fish are raised in turbid water and can rarely be seen. The best indication that fish are sick or stressed is a drop in feeding activity. Other factors that will cause this behavior are a drop in temperature, low oxygen, high carbon dioxide, and excessive activity around the pond. Of course feeding activity is only observable in supplementally fed ponds.

Other indications of disease are unusual surface activity or swimming patterns by the fish. They may swim in spirals, or "cork screw" or swim erratically, up and down. A common swimming pattern associated with skin parasites is "flashing". The fish rub their sides against the pond bottom or submerged objects and their white ventral surface flashes. Fish in respiratory distress gather around inflowing water where more oxygen is usually available. Weakened fish can be carried by the current and impinged against screens or drains. A prevalence of predator and scavenger animals around the ponds (they catch diseased dying fish) and the presence of sick or dead fish also indicate a disease problem. It is best to observe the fish twice a day. Dead and dying fish are often easier to find early in the morning before scavengers get them. High temperatures and high pH can affect fish in the afternoon and make stress due to ammonia more prominent. Finally, lesions and deformities are sometimes noticed when fish are harvested or seen when fish are feeding. Experience will help the farmer recognize problems. The farmer should be aware of the natural appearance of the fish in the pond at maturity and during spawning season. Often natural pigmentation spots or spawning tubercles in males of cyprinid (Minnows and carps) and catastomid (suckers) fish are mistaken for lesions. If fish do not consume enough food, they will emaciate, or waste away. In this condition, the head is much wider than the body, so the fish are called "pin heads".

The most common clinical signs of disease are ascites, exopthalmia, lepidorthosis and hemorrhage. Ascites is the accumulation of body fluid in the body cavity and is commonly called "pot belly". This fluid moves freely when the fish is handled and should be easily distinguished from a firm, full, distended stomach. Lepidorthosis occurs when fluid accumulates in the scale pockets beneath the scales. The scales protrude and give a "pine cone" appearance. Fluid accumulation in the eye socket pushes the eyeball out, resulting in exopthalmia or "pop eye". When blood escapes vessels, areas of hemorrhage occur and appear as red spots or reddened areas. Petechiae appear as small (pin point, less than 2 mm in diameter) red dots. They commonly occur in bacterial and viral infections. Ecchymotic hemorrhages are 3 mm or larger in diameter and typically occur at fin bases and around the anus during bacterial and viral infections. Ecchymotic hemorrhage is also frequently associated with lesions.

Gross indications of disease

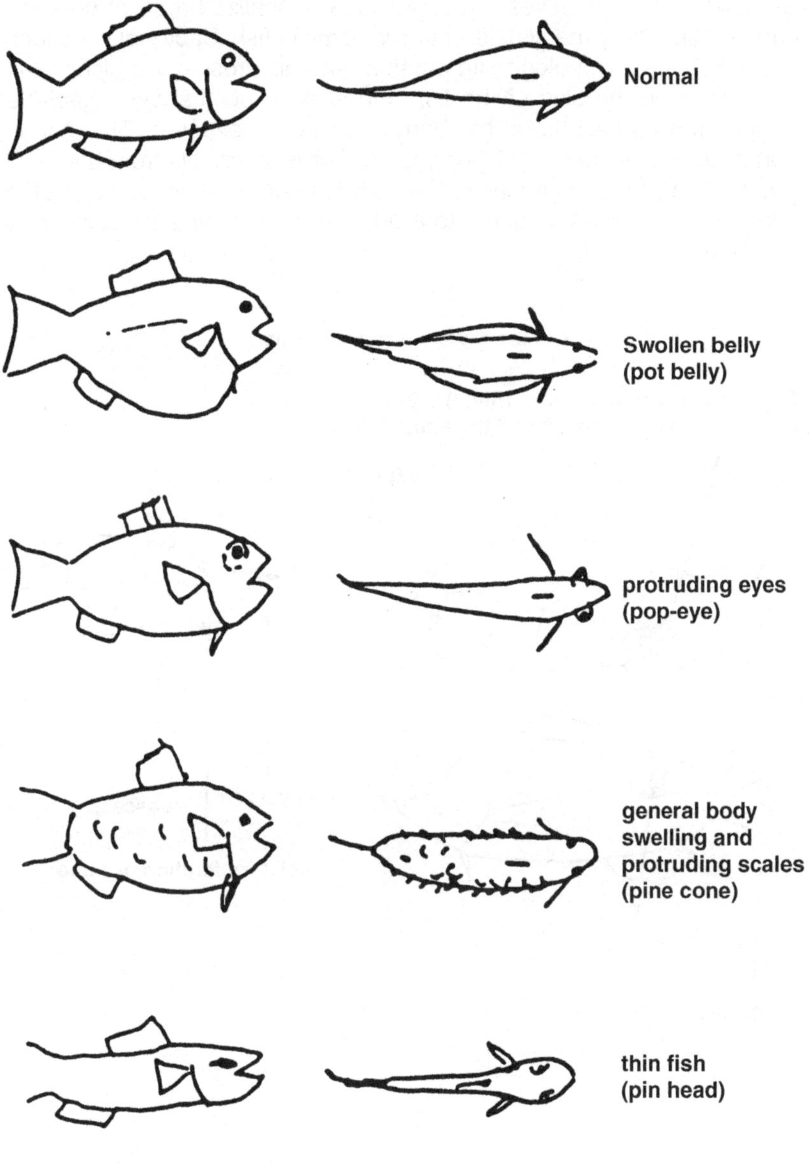

A cyst is a sac enclosing fluid or matter. In fish, cysts are typically associated with parasites. An abscess is a localized area of pus and can be caused by many different organisms in fish. Bacterial infections typically cause skin ulcers, fin erosion, and gill erosion. An ulcer is an open sore on the skin of the fish. Fin erosion occurs when bacteria destroy the fin tissue and the bony fin rays are exposed. This condition is commonly called "frayed fins". Gill erosion results from bacterial destruction of the gill lamellae. The gill appears to have pieces eaten away. The gill tissue adjacent to eroded areas is typically uneven and discolored (not red).

Commonly seen lesions in fish

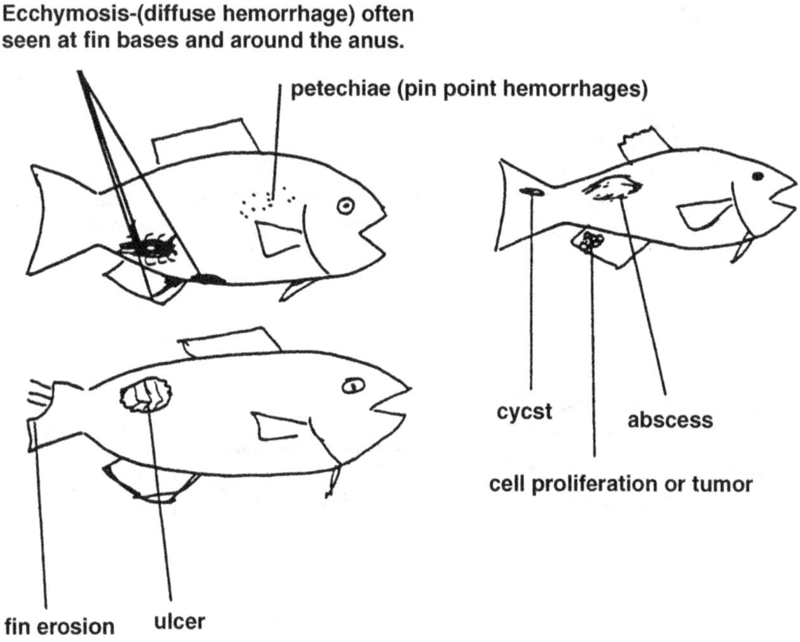

Common gill lesions

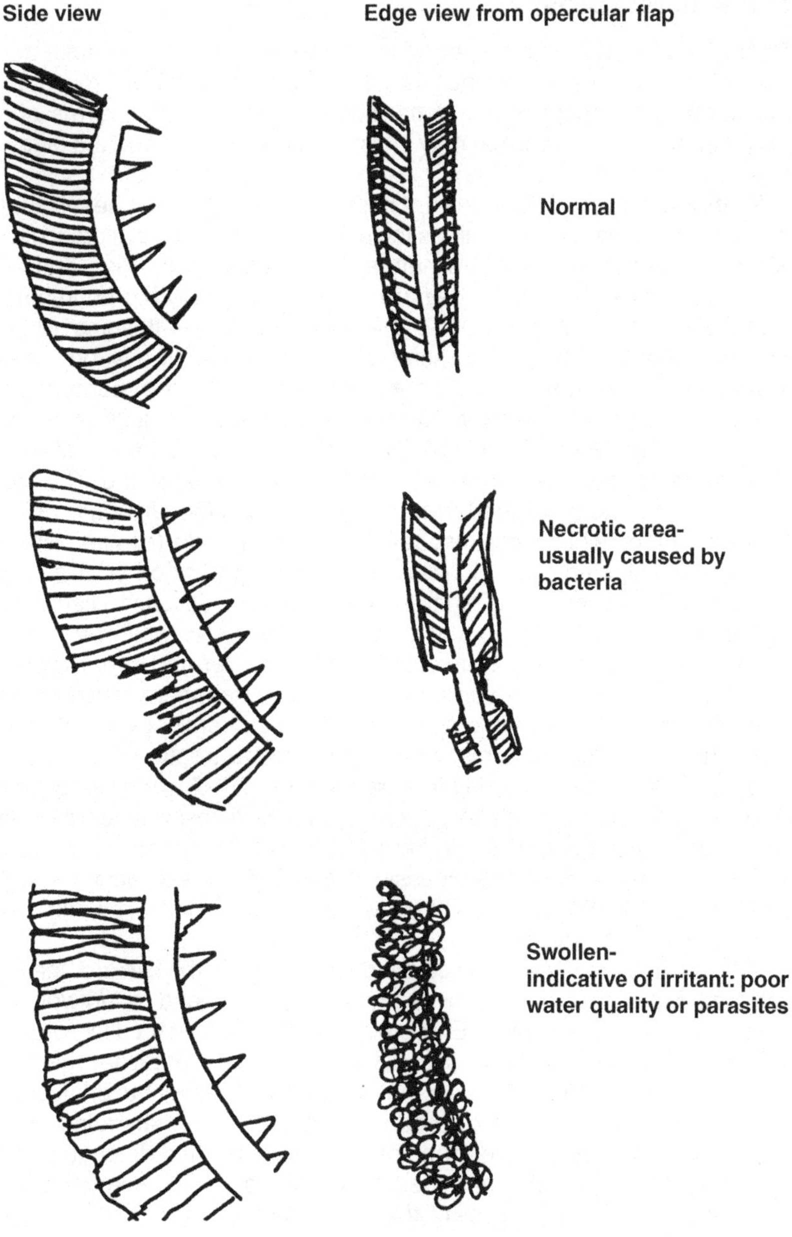

PATHOGENS

Parasitic Diseases

Parasitic infestation can be very stressful to fish, can cause lethal diseases and can affect meat quality. Parasitic protozoans (microscopic) and metazoans (especially worms such as flukes, tape worms and nematodes or round worms) may infest the gills, skin, blood, and digestive tract.

Protozoan caused diseases are best diagnosed by a fish health specialist (if available) who examines the skin and gills under a microscope. Most protozoan parasites are natural inhabitants on the fish and do not cause problems unless they are on the fish in very high numbers. Signs of infection with parasites are those of irritation—"flashing" (darting movements with the fish scraping its side against a submerged surface) or sometimes jumping out of the water. Fish may congregate at the surface or in shallow water, have an excess of mucus (seen as a gray or blue color), or they may crowd around an inflowing stream of water. Also feeding activity will almost always be reduced. The two common parasites that lead to mortalities are *Ichthyophthirius*, which causes 'Ich' or white spot disease, and *Ichthyobodo* (costia), which causes blue slime disease. Ich is caused by a large spherical ciliate up to 1mm diameter with a "U" shaped macronucleus. Diseased fish have extreme mucus production and small 1mm diameter white bumps in the skin. This parasite often infects wild fish so river and stream water sources can be the reservoir for this problem. Treatment requires the use of malachite green in locations where this chemical is legal or at least three successive formalin or copper sulfate treatments at 2 day intervals if over 20°C or 3 day intervals if under 20°C. *Ichthyobodo* is a very small bean shaped flagellate (7–10 mm diameter) with two flagella that causes extreme irritation to fresh water fish. The chemicals listed at the end of this section are successful in treating these diseases. There are a variety of other protozoa that are commonly found on fish and may occasionally case disease.

Metazoan parasites rarely cause disease unless the fish are heavily infested. More often the parasites (especially flukes and round worms) affect product acceptability. Encysted fluke larvae (often called grubs) occur in the flesh of many species of fish. These parasites have a series of hosts in their life cycle that consist of the adult fluke infecting a predator such as a bird or predatory fish, and then shedding eggs in the digestive tract of the predator. The predator then defecates into the pond, the eggs hatch and the larvae infect a snail, the larvae grow then leave the snail and infect the fish. Infected fish are then eaten by another predator in which the fluke matures and initiates another cycle. Management of these parasites involves breaking the life cycle. Bird numbers can be reduced by removing structures that birds roost on (especially those that overhang the pond) and by using various methods of scaring them away. Additionally, stocking fish that consume snails such as the black carp will help interrupt the life cycle. Large larval round worms also commonly infect the flesh of fish and reduce marketability (these worms are often red and migrate through the flesh after the fish is processed). They generally infect predatory birds as the final host and can be managed much like the flukes. **Caution:** several larval flukes, roundworms and tape worms that infect fish can infect humans and thus all fish flesh should be cooked well before consumption.

External metazoan parasites include gill and skin flukes (monogenetic trematodes) and copepods. The gill flukes are very tiny parasites 1–2 mm long that infest the gills and skin. In high numbers they cause similar problems as the external protozoan parasites. The most important of the copepod parasites are the anchor worm and the fish louse. The anchor worm is 0.5 to 2 cm long. It is strongly anchored in the fish's skin by burying a structure found in the anterior portion of the worm. This parasite is a common problem with scaly fish such as carp. Fish lice are shorter (0.3–0.75 cm) and hold on to the skin via hooked feet. Heavy loads cause skin irritation and bacterial infections. These external parasites have only the fish host in their life cycles but because they are external, they are accessible to external treatments. The most effective treatment for most external metazoan parasites is Masoten or Dylox (trichlorfon). This chemical can be toxic to fish and, if not handled properly, to humans.

Bacterial Diseases

All ponds carry a population of bacteria that can infect fish and cause disease. When fish are healthy, these bacteria may not harm them. If the fish are stressed, they have less resistance to infection and a bacterial disease can quickly spread through a population. Most bacteria that cause disease in fish are gram negative organisms. They can be identified by culture and staining in a laboratory. If fish are showing the following signs, the farmer should begin to suspect bacterial infections: skin lesions (ulcers), fin erosion, pot belly, exophthalmia (pop-eye), bloody spots on the skin, reduced feeding activity, erratic swimming behavior, discolored skin, fish grouped in shallow water, or gradual or sudden increase in mortality rate.

If it is possible, a farmer should take a few of the affected fish and some water to a local expert on fish diseases. Always take the fish which are still alive and which are representative of the kind of infection you are seeing. There are several things the farmer can do himself to help his fish. Increasing water flow and aeration may help in acute infections if the water quality is causing the stress that brought on the disease. If the fish are still feeding, it is best to continue feeding, but at a reduced rate. If supplemental feed is being given, an antibiotic may be incorporated into the feed. Although antibiotics are not a potential treatment in some areas either for financial reasons or because of the lack of supply, there are areas where these agents are available. One of the most commonly available antimicrobials for fish is oxytetracycline. Oxolinic acid is also widely used. Both of these drugs have a broad spectrum of activity against a wide range of pathogens. A few pages have been included at the end of this section listing the known dose rates of oxytetracycline and oxolinic acid for commonly farmed fish. An easy and convenient method of incorporating the drug in the feed is also given.

Restrictions on the use of antibiotics for use in food animals vary from country to country. Before any treatment is made, check with university officials or the proper government agencies for assistance with rates and application.

Viral Infections

Viruses cause diseases in fish. Most viral diseases are difficult to diagnose. Affected fish from severe outbreaks display general signs of stress as indicated by abnormal pigmentation (darker or lighter than normal) and lethargy (slow weakened swimming). Some viral diseases cause infected fish to have a pot-bellied appearance, they may swim erratically, and have bulging eyes (exophthalmia). Frequently, viral infections in the fry and fingerlings may cause high mortality. In some cases the surviving fish may show poor growth rates or develop with deformities. Viral infections may therefore result in a loss of profit for the farmer. Other virus infections cause unsightly bumps on the skin of the fish that affect marketability. These infections, like parasitic and bacterial infections, are frequently seen in conditions of crowding, poor water quality, and very high stress factors in the environment. Depending on the type of virus, if a pond is infected, the farmer may decide to kill all his fish, dry out the pond, and disinfect the system and start over. Viral diseases can often be prevented using avoidance techniques. If there is a known virus that causes problems in the fish you are raising, purchase fry or eggs from certified virus free sources. Do not import fish from areas with known virus problems and avoid mixing stocks of fish. Survivors of virus disease outbreaks should not be used for brood fish because many viruses are transmitted from parent to offspring through the eggs. Farmers initiating a carp culture system should avoid sources with spring viremia of carp virus and grass carp reovirus.

Fungal Infections

Fungi may adhere to the skin, fins, and gills or may be associated with systemic (internal) infection. These organisms are usually associated with external gray-white patches that resemble cotton or wool. Fungal infestations may also adhere to egg masses. The fungal matte can rapidly spread and suffocate the egg mass. Chemical treatments (potassium permanganate, formalin, and salt) have been used for fungal infections but relief is usually just temporary. Generally fungal infections are a sign of weakened condition of the fish during cool temperatures or an attempt to hatch eggs in water of poor quality or water at too cool of temperature. Fungal problems often recur in ponds with low alkalinity or hardness and if the pond is approaching the minimum temperature for survival for the species of cultured fish. A rapid drop in temperature could also cause the fish to be more susceptible. In areas exposed to rapid winter temperature changes the ponds should be maintained full of water to slow the rate of change.

General health management principles for fish carrying any of these infections are similar to those for controlling infection in other animals:

1. Always try to disinfect nets and equipment that are used in more than one pond to help prevent spread of infection.
2. A farmer should not walk into an infected pond and then attend to his healthy fish.
3. Careful monitoring of water quality as outlined earlier in this section.
4. Drying ponds and equipment thoroughly will help prevent many disease problems.

Therapy

Proper treatment of diseased fish is a delicate operation that requires considerable expertise. Improper treatment will generally disturb the balance in the pond, stress the fish and cost the farmer substantial amounts of money. Some chemical treatments kill phytoplankton and cause oxygen depletions. Also stress due to exposure to an external parasite treatment when fish are dying from a bacterial disease will often make the losses much worse. Indiscriminate use of antibiotics increases chances that the bacteria in the pond are resistant to the drugs which would make the drugs useless when it is really needed. The farmer should also be aware of the cost vs. benefit of treating. Generally, fish diseases are much like those of humans, the fish become diseased for 10–14 days then the immune system takes over to help the fish recover. Proper treatment will increase the numbers that survive and the effectiveness depends on how early in the progression of the disease it was noticed, diagnosed and treated. Often the farmer is better off managing the pond so the environment is not stressful and removing dead fish than to spend the money on a treatment. Finally, most chemicals are hazardous to the farmer and the fish if care is not taken in calculating the dosage, handling and applying the chemicals. Before an unfamiliar treatment is carried out in a new system, or with a new species of fish, a bioassay should be performed on a few fish. This is done by taking a few fish from the infected pond and placing them in a smaller container (being careful to make sure they have plenty of oxygen). The drug of choice for treatment is given at the correct concentration. The fish are observed for signs of stress. Once the treatment is completed, these fish are transferred to fresh water and watched for a further 12–24 hours if possible, for signs of delayed mortality. If these fish tolerate the treatment well, the rest of the affected fish may be treated. An experienced field worker should always supervise a treatment schedule. Chemicals used should be of good quality. Accuracy of treatment is very important. An overdose can kill all the fish in a pond, so calculations for dosage should always be double-checked by a second person.

Care should also be taken with the possible effect that the chemicals may have on aquatic systems down stream if you are treating a flow-through system. Sometimes it is difficult to find a sensitive scale (which may be required to treat small ponds and tanks). A concentrated solution of the drug (100 times the desired strength) can be made in water. Then, one hundredth of this solution (after thorough mixing) can be applied to the pond to be treated.

The effort spent in preventing diseases is always better than trying to treat the disease after it has occurred. Good preventive measures on fish farms include drying and liming of ponds between use and constant attention to water quality. It is better to stock a pond or cage with fish, harvest all the fish, clean up the system, and start over, rather than continually stocking and harvesting a pond. Dead fish should always be removed from the pond and buried and/or put into lime pits rather than left in the open. Regular flushing of tanks with fresh water also helps prevent disease. Species of fish vary in their susceptibility to the stress of crowding and handling.

External chemical treatments are given by dissolving the chemical in the water. Generally, solid chemicals such as potassium permanganate or copper sulfate must be dissolved completely before adding them to the pond or tank. The four types of external chemical treatments are administered as dips, flushes, baths or indefinite exposures.

Dip

Fish are placed in a net and dipped in a strong solution of a chemical for a short time. This can be very dangerous as the chemicals are in a concentrated form. Also this is very stressful to the fish. Generally dip treatments are not used on diseased fish but are instead used to remove external parasites from fish before stocking.

Flush

A known amount of a chemical solution is added to the upper end of a flow-through system and allowed to flush through the unit. This cannot be done in ponds and must be used in a system with adequate water flow, such as tanks and raceways.

Baths

The calculated amount of the chemical is added to the tank and allowed to stand, usually for 30–60 minutes, then flushed out of the system. Aeration should be given for all methods, but especially with baths. If fish show signs of distress (piping at the surface, erratic behavior, etc.), fresh water should be added.

Indefinite

The indefinite treatment is a common method used in ponds. A low concentration is applied to the pond and allowed to degrade naturally. Care should be taken to avoid uneven concentrations in localized areas.

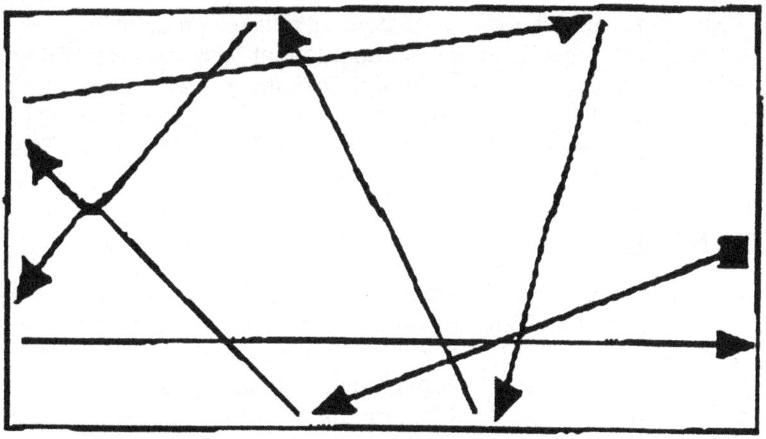

application pattern on ponds to prevent hot spot formation

When applying the chemicals to a large area (such as a pond), evenly distribute the chemical in a crisscross manner to prevent uneven concentrations of chemical in a pond. In small or narrow bodies of water the chemical can be dispersed from the bank using a sprayer. If you are applying chemicals other than salt or agricultural lime, make sure the chemical is dispersed in water before applying. Most chemicals take time to dissolve and will sink to the bottom and be lost in the mud if not first dissolved.

Calculation Of Treatment Levels

The correct volume of water to be treated must be known before attempting to calculate the amount of chemical to be used.

Volume of a pond = LENGTH X DEPTH X WIDTH

The units may vary (cubic meters, acres, hectares) but they must remain constant throughout the calculation. The farmer may already know the volume in gallons or liters. Use table 16 to help with conversion factors for all the units.

When the pond volume is known, the precise amount of chemical needed can be determined by:

V X CF X PPM Needed X 100/% A.I.

V = Volume of water to be treated

CF = Conversion factor (weight of chemical to be used to produce one ppm in one unit of volume of water—see table 16).

PPM NEEDED = Desired concentration of chemical in mg/liter or g/1,000 liters.

100/% A.I = 100 divided by the percent active ingredient.

Many drugs are sold in bulking agents that allow reliable mixing of small quantities of the drug or impart desirable physical properties for the use of the drug. For example, the drug you buy may have 1 gram of pure drug for every 5 grams of powder (1/5 x 100% = 20%). This is then called a 20% active preparation.

EXAMPLE:

How much potassium permanganate is needed to treat a pond that has an area of 100 m x 100 m and an average depth of one meter?

a) Volume of pond = 100 m x 100 m x 1 m = 10,000 m^3.
b) Conversion factor = 1 gram per cubic meter = 1 ppm.
c) Desired concentration = 2 ppm (dose which is needed).
d) Potassium permanganate is 100% active 100/100 = 1.
1 g = 1,000 mg

*See Table 5.
V X CF X PPM Needed X 100/100 = Amount chemical needed.
10,000 cubic meters X 1 g/cubic meter/ppm X 2 PPM X 1 = 20,000 g
= 20 kg Potassium Permanganate

Chemicals Commonly Used For Fish

The farmer should always obtain advice about the use of any compounds in his local area. He should also **be aware of the government laws and restrictions concerning the use of each drug or chemical.**

1. **Povidone—Iodine Solution** (Betadine)—equipment disinfectant. Is also used on eggs as a bath of 10 to 100 ppm for 10 minutes on one to two day old eggs.

2. **Salt** (sodium chloride)—general use therapeutant. Used when transporting freshwater fish as a 0.5% solution. May be used as a dip treatment at 3% for a short period of time. Salt is also used to prevent brown blood disease (methemeglobinemia, resulting from high water nitrite levels). It is distributed in the pond at a rate to give a ratio of chloride to nitrite of 3:1 or greater. Also, salt improves the ability of fresh water fish to handle stressful conditions and can be added to a pond at concentrations up to 1000ppm (1ppt).

3. **Copper sulfate**—100% active—effectively controls many algae. Copper sulfate is often used for parasites, primarily protozoans such as ***Trichodina***, ***Ichthyobodo***, ***Trichophyra***, ***Ambibphrya***, and ***Ichthyophthirius***. It is toxic in low alkalinity water (less than 50 ppm). The treatment rate often used is 0.01 ppm/ 1ppm alkalinity. The treatment for a pond with an alkalinity of 75 ppm would be 0.75 ppm copper sulfate. This agent must be dissolved well before dispersal and can kill the algae, resulting in oxygen depletion. This agent is very corrosive to aluminum i.e., boats and containers.

4. **Formalin** (37% formaldehyde) this solution is considered 100% AI in calculations—Used for ectoparasites as above. An indefinite pond treatment of 15–25 ppm is recommended. Formalin is toxic to man and causes respiratory and skin irritation. Formalin splashed into the eyes can cause blindness, so great care should be taken when using this chemical. In warm water, formalin will reduce the oxygen content so aeration should be available. If formalin is stored too long or stored at <45F it degrades and forms a white precipitate which is paraformaldehyde and is toxic to fish.

5. **Dylox** (Masoten), or Trichlorfon—Usually obtained as an 80% powder (80% A.I.). This chemical effectively controls leeches, monogenetic trematodes (gill worms), and anchor parasites and is applied at 0.25 ppm active ingredient. It should not be used at a pH of 8 or higher. In ponds, the pH is higher in the afternoons, so it is best to treat in the

morning. **This is an organophosphate pesticide and is toxic to man and animals. Sensitivity of fish species varies greatly.**

6. **Potassium permanganate**—100% active ingredient. This chemical is effective against ectoparasites and external bacteria at 2 ppm above the organic permanganate demand. When added to a pond, potassium permanganate will first oxidize the organic matter suspended in the pond water and then oxidizes protozoan parasites, bacteria and fungus on the skin of the fish. The chemical remains a wine or red color while it has oxidizing potential, and turns orange or brown when it is no longer effective. The water must stay a red color and not turn brown for at least 12 hours for an effective pond treatment. If the water turns brown, additional chemical is needed and should be added at increments of 1 ppm. The color change is difficult to perceive in many ponds, so more than 6 ppm should not be added within a 24 hour period. **Potassium permanganate is a potent oxidizer and will destroy clothes, stain the skin and damage eyes. This agent can cause spontaneous combustion if it comes in contact with oil or grease.**

7. **Malachite green**—100% active powder. Only zinc free product should be used on fish. This chemical is effective against parasites and fungus when added to the water at 0.1 to 0.2 ppm. It is very toxic to some species of fish and has a narrow safety margin. It is not legal for use on food fish in the United States of America because it is **a suspected cancer causing agent and could cause mutations**, but it is legal in many other countries. Because of the potential human health risk of consuming this chemical the use of this chemical should be restricted to fingerlings, fry and eggs. This chemical is a dye and is usually sold as a fine powder. Care must be taken when working with the powder so that it is not inhaled.

Feeding Drugs

Medications with systemic sites of action are often administered as feed additives. This allows the treatment of a large portion of the fish population without the labor needed to inject fish or the extreme expense needed to apply therapeutic doses in the water. The limitations of using medicated feeds are

—the therapeutant must be effectively absorbed through the GI tract or have a GI site of action

—the fish to be treated must be actively eating supplemented feed at the time

—because of individual variation in feeding, the treatment is not evenly distributed—some fish receive a high dose and some receive no treatment.

Medicated feed is commercially available in some areas where intensive aquaculture is practiced. These feeds are formulated for specific feeding rates and are generally manufactured with the medication incorporated into the feed particle. This type of formulation provides the most reliable therapeutic application because the medication does not leach out of the feed significantly before the particle is eaten. Where medicated feed is not readily available the medication can be mixed into moist dough balls or suspended in vegetable or animal fat or gelatin (1:32 gelatin powder to water) and coated on the outside of the feed. To assure that the fish receive the medication before it leaches out of the feed, apply a small amount to the pond to attract the fish, until the fish are actively feeding then apply the normal amount. Some therapeutants have a repulsive taste and cause the fish to reject the feed. This feed can be made more palatable by coating it with a small amount of fish oil. Medicated feed should not be stored for long periods. The therapeutant, vitamins and lipids will degrade resulting in a compound that could worsen the disease problem. Medications provided in the feed are dosed according to the amount of active ingredient to body weight of fish. Therefore the feeding rate (% of body weight /day) is factored into the treatment calculation.

Dose = amount of active ingredient/ weight of fish

$$\frac{\text{premix}}{\text{weight of feed}} = \frac{\text{Dose}}{\text{fraction of BW feeding} \times \text{fraction of premix AI}}$$

For example: If you were to treat tilapia that were eating 3% of their body weight per day with Romet 30 at a dose of 50 mg/ Kg fish and the premix was 33g AI/100 g. You would need 5 g of premix per Kg of feed.

50mg = 0.05g

$$\frac{0.05\text{g/Kg fish}}{(0.03\text{ Kg feed/ Kg fish}) \times (33\text{g AI/ 100 g premix})} = \frac{0.05\text{ g/Kg} = 5\text{g premix/ Kg feed}}{0.01}$$

Injections

This method may be useful for individually sick brood fish. Care must be taken when handling these fish since they are already stressed because of infection. Injections are either given intramuscularly (IM) or intraperitoneally (IP).

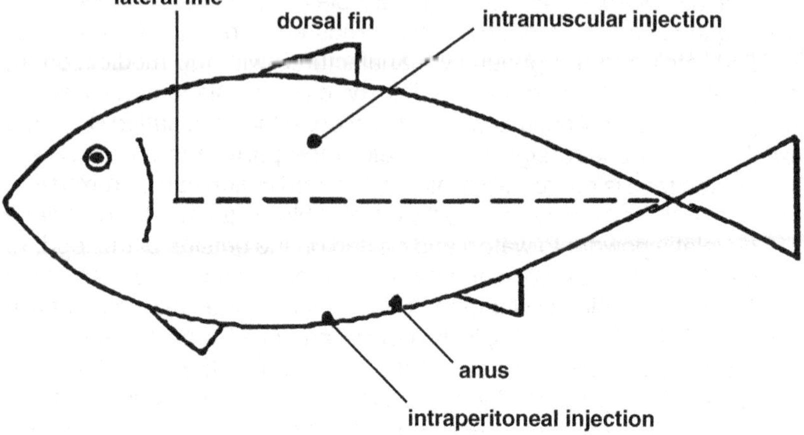

Injection sites for fish

Antibiotics Commonly Used For Fish

1. **Oxytetracycline (Terramycin)**—A broad spectrum antibiotic for the control of bacterial infections. Check with local authorities for information on the use of antibiotics in food animals.

Treatment options available:

BATH—15–20 ppm active for 24 hours repeated daily for 10 days.

INJECTION—IP or IM at 50 mg per Kg body weight daily for 10 days.

FEED—60 to 75 mg active drug per Kg fish per day for 10 days.

Oxytetracycline has a recommended withdrawal time of 21 days in the USA. That is, the fish must not be killed and sold for food until at least 21 days after the last treatment of the drug.

2. **Ormetoprim + Sulfadimethoxine (Romet)**—30% dose = 50 mg active per kilogram of body weight per day for five days. In the United States, withdrawal time is 45 days for fish that are sold with the skin and 3 days for fish that are sold without the skin.

3. **Trimethoprim + Sulfadiazine (Tribrissen)**—dose = 30 mg active per kg body weight per day for five to seven days. Withdrawal time is 21 days in the United Kingdom.

4. **Oxolinic acid** (Aqualinic Veterinary Powder)—dose = 10 mg active per kg body weight per day for 10 days. Withdrawal time is 30 days in the United Kingdom.

Producers should consult local authorities for information regarding the use, restrictions and availability of therapeutants in their region.

Section VII

Zoonotic Agents, Food Safety and OIE Regulations

Fish are almost always safe to handle and eat. Since they are cold-blooded animals, most pathogens that infect fish will not live in a man's body. However, there are some organisms that occur in some fish that can be transmitted to man, particularly people that are very young, very old or sick. Although rare, the following bacteria, parasites and toxins have been reported to cause disease in humans. Anyone handling or consuming aquaculture species should be made aware of these potential hazards. If illness does occur, the attending physician should be made aware of potential pathogens carried by fish. If fish are sick and/or dying, consult Section 6: Fish Health Management.

Bacteria

Erysipelothrix is probably the most common disease transmitted from fish to man. These bacteria are transmitted to man by many animals. This disease commonly occurs in butchers and veterinarians. It does not cause disease in fish or animals. The most common form of this disease in man is a skin infection. Septicemia, or blood poisoning has been reported, but rarely occurs.

Streptococcus species have been reported in many different types of fish. It most commonly occurs in tilapia, striped bass, ayu, yellowtail, rainbow trout and eels and affects aquaculture in North America, South America, Asia, Japan and the Middle East. *Streptococcus* can be transmitted to humans and can result in septicemia, particularly in the very young, old or sick. If fish are sick with *Streptococcus*, generalized signs of a bacterial infection are present (see Section 6). Signs that are more specific to this disease include swollen, infected eyes and a rolling, spiraling swimming pattern, resulting from bacteria in the brain. Humans are infected when placing their hands in the culture water, or handling or cleaning infected fish. Bacteria can enter through breaks in the skin or in cuts received while handling or cleaning the fish. Usually the infection remains localized but can become systemic. Treatment is usually a topical antibiotic. An oral antibiotic can be administered if needed.

Mycobacterium occurs in marine and freshwater fish worldwide. Most fish are potential hosts, but the most severe disease and greatest potential for transmission to humans occurs in cultured striped bass and striped bass hybrids. *Mycobacterium* usually occurs as a chronic disease and the generalized signs of a bacterial infection are usually present. In the final stages, fish may demonstrate protrusions or bumps. These result from internal nodules caused by the disease. Humans are infected when placing their hands in the culture water or handling or cleaning infected fish. Bacteria can enter through breaks in the skin or in cuts received while handling or cleaning the fish. In humans,

Mycobacterium is usually restricted to the fingers and hands in the form of a chronic granulomatous (hard noduled) infection. Treatment is usually a topical or oral antibiotic. Surgery to remove granulomas may be required in severe cases. Additionally, *Mycobacterium* can cause respiratory disease or arthritis in people with a weakened immune system.

Vibrio usually occurs in saltwater fish and shellfish but can occur in fresh water species. There are many species that infect fish and shellfish, but only 3 of these have been reported to cause disease in humans. In fish, *Vibrio* infections result in a hemorrhagic septicemia. In humans, infection is usually associated with shellfish and is associated with eating raw or undercooked products. If the bacteria are eaten, symptoms range from mild to severe vomiting and diarrhea. Skin infections result if the bacteria enter through a break in the skin. A rarely reported type of *Vibrio* can result in bacterial septicemia (blood poisoning) and can result in death.

Edwardsiella occurs in many species of amphibians and fish, resulting in generalized septicemia with the secondary formation of abscesses. This disease can be especially devastating in cultured eels. Humans can become infected through breaks in the skin or eating raw or undercooked products, resulting in localized skin infections or diarrhea, respectively. In people with a weakened immune system, or those already sick, death can occur. Treatment is a topical or oral antibiotic.

Aeromonas infections can occur in all species of fish, but these bacteria rarely infect humans. If this happens, a localized infection occurs where the bacteria enter through a break in the skin. Localized infections can progress to septicemia in people with a weakened immune system. Treatment is a topical or oral antibiotic. Aeromonas can be ingested in contaminated drinking water, causing diarrhea.

Plesiomonas occurs in fish held in intensive systems, but usually only occurs when fish have been crowded or stressed, or are suffering from another disease. It also can occur in shellfish. Humans become infected when eating raw or undercooked fish or shellfish, resulting in vomiting and diarrhea. This disease is usually not seen in healthy people. Most reported infections have occurred in people already suffering from a serious illness.

Parasites

Fish parasites that also infect man have many types of life cycles. This means that there are many developmental stages of the parasite and the different stages occur in different animals. A common life cycle of fish parasites involves small microscopic water organisms called zooplankton, snails, fish and a land animal. The parasites that can infect humans usually get into the water by human waste (feces), which carries the eggs. After the egg hatches, the zooplankton and/or snail is infected. When the fish eats the zooplankton or the snail, the fish becomes infected. If infected fish is eaten raw, or food is prepared on a surface where infected fish have been cleaned, the larval stage can infect humans and migrate through the body. The larval parasite then develops into another larval stage, or can develop into the adult stage. If the adult stage lives in the human gut, parasite eggs are shed in human waste.

Digeneans or flukes include members of the families Opisthorchidae and Heterophyidae. *Opisthorchis* species inhabit the bile ducts of man. This disease occurs throughout Asia. It can cause symptoms ranging from diarrhea to acute pancreatic disease. This parasite involves freshwater fish, especially carp. *Heterophyes* species occur in the Middle East, Asia, and the Baltic, in both fresh and salt water. This parasite is carried by Mullet and develops into an intestinal parasite in man.

Tapeworms include *Diphyllobothrium*. Fish are the intermediate host; man is the final host. The stage that infects man is found in the liver and muscle of infected fish. This tapeworm occurs in freshwater fish in North America, Europe, and Asia.

Several nematodes or worms that infect man are common. *Anisakis, Contracaecum and Pseudoterranova* infest the stomach wall of people and cause chronic inflammatory lesions. Larvae also migrate to the muscle, causing significant damage and pain. In the Far East and Pacific, fresh water and marine fish can carry *Angiostrongylus*. If ingested, these worms can cause inflammation of the brain (meningitis). In the Philippine Islands, a *Capillaria* worm causes a severe form of diarrhea. Eggs are passed through human feces in the water, fish are infected and then infected fish are eaten raw.

In many countries, human excrement, called night soil, is used as fertilizer on agricultural crops and in aquaculture ponds. The night soil can be treated or used untreated. The use of untreated night soil is not recommended. It usually results in transmitting parasitic diseases and diarrheas. Treated night soil does not cause diseases and has been shown to result in the production of better quality fish. To treat night soil, it can be placed in an anaerobic digester, or composted in a traditional composting manner. If the water is used for crop irrigation, minimal wastewater treatment may reduce bacteria, but not parasites. To reduce the spread of parasitic diseases, a two-stage waste water treatment is required.

Several things can be done to reduce bacterial and parasitic infections in man. Human waste or feces should not be disposed of in areas where it will contaminate drinking and swimming water. If a water source is known to contain human waste, do not swim there and boil the water before using it for drinking or cooking. Thoroughly cook fish before eating it and clean all cooking surfaces and utensils with fresh water.

Toxins

There are also food-born toxins associated with eating fish. In tropical coastal regions, the ingestion of large predatory reef fish can result in ciguatera poisoning. Algae produce the ciguatoxin. Small reef fish eat the algae. Larger fish eat the smaller fish and the toxin accumulates in the larger fish's flesh. When a human eats the affected fish (and the ciguatoxin), vomiting and diarrhea may develop. If high levels of the toxin are present, secondary nervous system problems can develop.

Another toxin associated with fish is scombrid poisoning. It usually occurs after eating tuna, mackerel, bonito or skipjack. Scombrid poisoning is a result of the early stages of decomposition in these fish, so they should only be eaten fresh. Symptoms include vomiting, diarrhea and hives.

Eating coastal shellfish may result in paralytic shellfish poisoning. Algae that the shellfish eat produce the toxin that causes this. In humans that ingest this toxin, extensive neurological damage can occur. However, most cases are not that severe. In many countries, shellfish populations are monitored and warnings are posted if detectable levels of the toxin are found.

Food Safety

Fish flesh can become contaminated with gut bacteria during cleaning or processing. *Clostridium* naturally occurs in the fish gut. These bacteria produce toxins that when eaten by man, result in vomiting, diarrhea, respiratory distress and muscle weakness. *Pseudomonas* and *Aeromonas* are also normally found in the fish gut and can contaminate fish during processing. The bacteria *Vibrio* is responsible for most sea food illnesses (referred to in the bacteria portion of this section).

Fish flesh can also become contaminated after cleaning or processing. Flies and rodents carry bacteria, or the fresh fish can be placed on a dirty surface or handled with dirty hands. The bacteria can rapidly multiply on the fish flesh. If the fish is not cleaned and cooked before eating, bacteria or toxins may be ingested and vomiting and diarrhea can result. If a high amount of bacteria is present, the fish appears obviously rotten and usually is not consumed.

The other bacteria that most commonly result in food-borne contamination illnesses are *Salmonella, Escherichia, Staphylococcus, Bacillus* and *Clostridium.* To prevent food-borne illnesses, fish must be processed quickly, rinsed well, then preferably placed on ice or kept chilled.

If ice is not available, fish should be held in a shaded area free of rodents and flies. People processing and handling the fish should keep their hands, utensils and working surfaces as clean and fly free as possible.

In most countries, some form of HACCP (Hazard Analysis Critical Control Point) has been adopted. This program is designed to identify points in the processing of a food where contamination is most likely to occur. Steps are then taken to prevent possible contamination.

Guidelines for Importing or Exporting Fish

Federal and local governments often have detailed regulations on the importation of live fish and shellfish. These regulations specify species and health status of the fish being imported and are designed to prevent the importation of undesirable or non-native species or a disease that is not present in the country. You must be aware of the regulations for your country before you import fish or eggs to your facility. Knowing other countries' regulations is important if you plan to market fish in these countries. Health status regulations are generally designed to protect a national resource without restricting free trade. The individual producer may want to stress more stringent health status for the fish being brought onto a facility. Diseases that are caused by pathogens that require hosts for survival (obligate pathogens) are limited in distribution and can be effectively controlled by avoidance. Purchasing specific pathogen free eggs or fingerlings provides for this. The protocols for inspecting lots of fish and certifying a facility disease free, a list of many of the pathogens that are of concern and a list of qualified laboratories that can inspect fish are provided by the Office International Des Epizooties (OIE), 12 rue de Prony, 75017 Paris, France. Telephone 33(0) 1 44151888, Fax 33 (0) 1 42670987, email oie@oie.int. The OIE also keeps track of the OIE countries that are free of specific diseases of concern and monitors the outbreaks of these diseases to update the status of these countries. The specific diseases that are monitored and reported if found (in OIE monitored countries) are termed 'notifiable diseases'. The following aquatic animal diseases are OIE notifiable:

Epizootic hematopoietic necrosis—a viral disease that has been associated with high losses of rainbow trout in Australia.

Infectious hematopoietic necrosis—a viral disease that causes high losses of salmonids in countries along the northern coasts of the Pacific Ocean. It has recently been introduced into Europe.

Onchorynchus masou virus—a viral disease that causes high losses in *O. masou* in East Asia.

Spring viremia of carp—a viral disease that affects carp, minnows, goldfish, tench and European catfish in Europe.

Viral hemorrhagic septicemia—a viral disease that affects salmonids and has also been associated with losses of herring, pike and white fish. The most virulent strain occurs in Europe. There is a form in the north Pacific that has been associated with losses of herring.

Bonamiasis—a protozoan disease of oysters in Australia, New Zealand, Europe and North America.

Haplosporidiosis—(also MSX and SSO)—a protozoan disease of oysters that is endemic to North America and has been reported in Korea.

Marteiliasis—a protozoan disease of oysters in Europe and Australia.

Mikrocytosis—a protozoan (*Mikrocytos mackini*) disease of oysters on the West coast of Canada and on the East Coast of Australia. This disease is also called Denman Island Disease.

Taura syndrome—a viral disease of penaeid shrimp in North and South America and Southeast Asia.

White spot disease—a viral disease of penaeid shrimp in Asia and North America.

Yellow head disease—a viral disease of penaeid shrimp in Asia, North and South America and Australia.

Classification of infectious diseases by species affected and possibility of control.

Aquaculture species	***Diseases likely to be controlled by avoidance***	***Diseases likely to be ubiquitous (but avoidance may be helpful)***
Cyprinids: carp, minnows and goldfish	Spring Viremia of Carp Carp Pox Grass Carp Reovirus Koi Herpesvirus Golden Shiner Virus Goldfish, Carp Ulcer Disease (*Aeromonas salmonicida*) Asian tapeworm Carp Gill Necrosis (iridovirus)	Pseudomoniasis Motile Aeromonas Septicemia columnaris External protozoan parasites monogenetic trematodes Saprolegniasis anchor worms nematodes and flukes
Cichlids: tilapia	Mycobacteriosis *Streptococcus iniae* Tilapia Iridovirus	Same as those in Cyprinids
Silurids: catfish	Channel Catfish Virus (Ictalurids) Enteric Septicemia of Catfish Proliferative Gill Disease European catfish virus	Same as those in Cyprinids Edwardsiellosis

(*continued*)

Classification of infectious diseases by species affected and possibility of control.

Aquaculture species	*Diseases likely to be controlled by avoidance*	*Diseases likely to be ubiquitous (but avoidance may be helpful)*
Gourami	Gourami Iridovirus	Same as those in Cyprinids
Crayfish (crawfish) and freshwater shrimp	Crayfish plague Baculovirus, microsporidians	ectoparasites FreshwaterVibriosis
marine fish	viral encephalopathy and retinopathy Infectious Pancreatic Necrosis Piscirickettsiosis Red Sea Bream iridovirus Photobacteriosis	Saltwater Vibriosis Saltwater columnaris external parasites Ichthyophonus
Penaeid shrimp	Yellow head disease White Spot Disease Taura Syndrome Baculoviruses Infectious Hypodermal and Hematopoietic Necrosis	Vibriosis ectoparasites

Section VIII

Economics

ECONOMICS

A farmer can begin to rear fish at a very simple level to provide food for his family or he may own several large ponds and sell his fish at local markets. Either system of operation requires a profit or gain from the venture. The small farmer will receive less reward for his labor but he may not have invested much effort or money in his pond. Although the large farmer may appear to be making a lot of money, he may use all of it in the cost of land, labor, and fish food. Generally, it is best to start off a new farming venture on a small scale. This is as true for fish farming as for other kinds of farming. As more experience is gained, the farmer can expand his production.

Fish farming can be an expensive and high risk enterprise. Higher stocking rates should result in higher yields and profit, but more capital is involved and there is an increased risk of fish kills due to oxygen depletion and stress-induced diseases. In areas where emergency aeration, well water and therapeutants are not readily available, moderate stocking, feeding and fertilizer rates should be used. The 'economics' of fish farming are difficult to standardize because the type of fish farming venture and the species used varies greatly. Also, the cost of land, water, electricity and feed vary from region to region and are always changing within a region. The type of enterprise will effect the economic evaluation. Is the desired product supplemental protein to be consumed by the producer, or is it a monetary profit? An important consideration in both cases involves the risk of fish farming: can you afford occasional losses, in dollars, effort or protein source?

If the fish will be harvested and consumed on the same farm, several points (and costs) to consider include

—land or pond site: if the land is not already owned, it must either be purchased or some type of secure holding must be obtained. Is the site suitable for pond construction or whatever type of venture has been chosen (i.e., cage culture)? Is the area protected from flooding?

—pond construction: cost of moving dirt and the equipment and labor involved; drains, pipes, valves and fittings, hand tools, plumbing supplies

—water supply: wells, pumps, fuel or electricity

—daily maintenance, observation or feeding: your time or the cost of hired labor

—emergency aeration, boat, nets, buckets, therapeutants

—fingerlings or stocker fish: must be produced, purchased or captured and transported

—food or fertilizer

—lost opportunity: the return on your effort, land and money if you invested it elsewhere

If fish farming is a business venture, and the fish will be harvested and sold, there are several other costs and factors to consider in addition to those mentioned above for subsistence fish farming:

—marketability: are there people willing and able to buy your fish, what is the investment into establishing a stable market, what is the cost of holding fish if the market is seasonal

—fish harvesting and hauling equipment and labor

—processing and storage

—interest on loans and taxes

Determine the cost of your initial investment to start the project by adding all of the above that apply to the type of fish farming you will be doing. The cost of all the items listed should be balanced against the gross profit expected. It is important that the farmer does not make a major investment in rearing large numbers of fish before he knows the problems he is likely to encounter. Also, because fish culture is a risky business the farmer should not over-extend his investment to the point that a loss of crop would jeopardize his farm.

A final point is that the overall key to good fish farming is good quality management by the farmer. It is very easy for a farmer to become unenthusiastic when he provides long hours of labor to a water system where he cannot see his fish. As his enthusiasm diminishes, his standards of care decline. He therefore produces a poor harvest and he will not try again another year. It is important to try and keep interest and excitement at a high level.

Section IX

Additional Information

ADDITIONAL INFORMATION

Since this booklet cannot cover all possible aspects of health and production practices in primitive fish farming. Christian Veterinary Mission is providing in this section a method whereby you may obtain additional information on aquaculture problems in your area.

The following is a general outline of information needed.

1. **Farm Location**
   a. Country
   b. Area of Country—State or province
   c. Distance and direction from nearest large city
2. **Climate by Season**
   Example: Winter, December—February, Cool (10°C) with heavy rains daily.
   Rest of the year—hot (30°C) and dry.
   Give seasonal extreme temperatures and amount of rain, if possible.
3. **Description of Area** where land is located
   Example—High mountain plateau, sparsely populated
   Example—Small farms, densely populated area near large city
   Example—range land with low swampy areas near stream
   Example—Tropical island with coconut and sugar cane plantations
4. **Description of Farm** in detail
   a. Terrain
      Example—Low, uncleared land with small amounts of well drained cropland being farmed
      Example—High mountain valley
   b. Total acres in farm
   c. Acres used for aquaculture
   d. List other land crops grown by season
   e. Acres used for other livestock
   f. Numbers and types of animals also on farm. Separate animals into immature, mature, and breeding stock
   g. Describe method of confinement, tethered, fenced, running lose
   h. Describe all buildings and sheds; age and condition
   i. Describe layout of aquaculture system, ponds, tanks, and cages. Draw a small map of the layout of the farm. Show water circulation throughout your system, especially where the water comes from and how the water drains away.
      Do you use your water supply for anything else—washing, water for other animals?

j. Describe distance to nearest farm with fish
k. Describe the numbers and types of fish on the farm, as well as stocking density.

5. Management Practices

a. Water Quality
Source—well, stream, pond
How supplied—
List any large farms or industries for ten miles upstream of your farm and what they are likely to discharge into the water.
Salinity temperatures extreme
Any water parameters you know of—hardness, alkalinity, pH. Do you leave ponds and tanks dry in between fish batches?
How long?

b. Nutrition
In detail
What foods are fed and to each age group
Seasonal variation
Include minerals and salts

c. Fish Supply
Do you buy fingerlings or do you rear your own?
Spawning—What system do you use? What time of year?

d. Treatments
List any drugs you use for treating infections.
List any chemicals you know are put on your land or adjacent land.
Which could get into your water supply?
How many years experience do you have in raising fish?

6. Describe Problem in Great Detail

a. How long has the problem been seen in the pond?
b. Does it have a seasonal pattern?
c. Are other fish in your local area sick?
d. What age fish are affected?
e. What percent of fish are affected by size?
f. What percent of affected fish die by size?
g. What is the average number of days from first sickness noticed until death?
h. Do survivors grow well or are they stunted?
i. Is spawning affected? Describe.
j. Is appetite affected? Describe.
k. Describe all symptoms in great detail.
l. Describe treatments used and results.

7. Any Other Observations? Remember, no detail is too trivial to be of great importance.

Mail to: Christian Veterinary Mission
19303 Fremont Avenue North
Seattle, Washington 98133 USA

TABLES

Table 1. Conversions for units of volume

TO *FROM*	*cm^3*	*liter*	*m^3*	*in^3*	*ft^3*	*fl. oz.*	*fl. pt.*	*ft. qt.*	*gal.*
cm^3	1	0.001	1x10^{-6}	0.0610	3.53x10^{-5}	0.0338	0.00211	0.00106	2.64x10^{4}
liter	1000	1	0.001	60.98	0.0353	33.81	2.113	1.057	0.2642
m^3	1x10^{6}	1000	1	6.1x10^{4}	35.31	3.38x10^{4}	2113	1057	264.2
in^3	16.39	0.0164	1.64x10^{-5}	1	5.79x10^{-4}	0.5541	0.0346	0.0173	0.0043
ft^3	2.83x10^{4}	28.32	0.0283	1728	1	957.5	59.84	29.92	7.481
fl. oz.	29.57	0.0296	2.96x10^{-5}	1.805	0.00104	1	0.0625	0.0313	0.0073
fl. pt.	473.2	0.4732	4.73x10^{-4}	28.88	0.0167	16	1	0.5000	0.1250
ft. qt.	946.4	0.9463	9.46x10^{-4}	57.75	0.0334	32	2	1	0.2500
gal.	3785	3.785	0.0038	231.0	0.1337	128	8	4	1

Table 2. Conversions for units of length

TO *FROM*	*cm*	*m*	*in.*	*ft.*	*yd.*
cm	1	0.01	0.3937	0.0328	0.0109
m	100	1	39.37	32.81	1.0936
in.	2.540	0.0254	1	0.0833	0.0278
ft.	30.48	0.3048	12	1	0.3333
yd.	91.44	0.9144	36	3	1

Table 3. Conversions for units of weight

TO *FROM*	*gm.*	*kg.*	*gr.*	*oz.*	*lb.*
gm.	1	0.001	15.43	0.0353	0.0022
kg.	1000	1	1.54x10^{4}	35.27	2.205
gr.	0.0648	6.48x10^{-5}	1	0.0023	1.43x10^{-4}
oz.	28.35	0.0284	437.5	1	0.0625
lb.	453.6	0.4536	7000	16	1

Reproduced by permission from Dr. Tom Wellborn

Table 4. Miscellaneous Conversion Factors

1 acre	= 0.405 Hectares
1 acre-foot	= 43,560 cubic feet
1 acre-foot	= 325.850 gallons
1 acre-foot of water	= 2,718,144 pounds
1 cubic-foot of water	= 62.4 pounds
1 gallon of water	= 8.34 pounds
1 gallon of water	= 3,785 grams
1 liter of water	= 1,000 grams
1 fluid ounce	= 29.57 grams
1 fluid ounce	= 1.043 ounces
1 Hectare	= 2.4711 acres
1 Chinese mou	= 0.1518 acres

Table 5. Weight of chemical that must be added to one unit volume of water to give one part per million (ppm)

Conversion factors	
2.72 pounds per acre foot	1 ppm
1,233 grams per acre foot	1 ppm
0.0283 grams per cubic foot	1 ppm
0.0000624 pounds per cubic foot	1 ppm
0.0038 grams per gallon	1 ppm
0.0584 grains per gallon	1 ppm
1 milligram per liter	1 ppm
0.001 gram per liter	1 ppm
1 gram per cubic meter	1 ppm
8.34 pounds per million gallons of water	1 ppm

Section X

Acknowledgments
Biography

ACKNOWLEDGMENTS

We have received information from many individuals and sources in preparing this book and we are very grateful for their assistance. If anyone or any source has been omitted, it is an oversight and we apologize. The publication of this book was made possible by donations from various individuals interested in the work of CHRISTIAN VETERINARY MISSION.

Special Acknowledgments

REVIEWERS

Dr. John Plumb, Dept of Fisheries and Allied Aquaculture, Auburn University, Alabama

Mr. Peter Kenworthy, Church Missionary Society, Tanzania.

Dr. Liang Zheng Fan, Guangxi Agricultural College, Nanning, China.

Dr. D.E. Goodman, College of Agricultural Sciences, Tuberville, South Carolina.

Dr. John Seamer, FRCVS Wiltshire, United Kingdom.

Dr. Bill Daniels, Wildlife and Fisheries, Mississippi State University.

Mr. Jim Collier, Extension Agent, Georgia

Dr. George Lewis, Extension Aquaculturist, Athens, GA

Dr. Ruth Floyd, Extension Aquaculturist, Gainesville, FL

Dr. Hank Stoddard, Clinic and Fisheries, Cross City, FL

Dr. Kelly Ward, Moscow, ID

Dr. Raymond Huml, Durham, NC

Sources of Information

1. FAO Technical Bulletins
2. Chakroff M., (1976), Fresh water fish pond culture and management. Peace Corps, U.S. Printing Office, Washington, DC 20402.
3. Tucker, C.S. (Editor), (1985), Channel Catfish Culture. Elsevier Science Publishers, B.V. Amsterdam, Netherlands.
4. Brown, E. and Gratzek, J.B., (1980), Fish Farming Handbook VI, Publishing Co., Inc., Westport, Connecticut.
5. Stickney, R.R., (1979), Principles of Warmwater Aquaculture, John Wiley and Sons, New York.
6. Pullin, R.S.V., and Shehadeh, Z.H., (Editors) (1980). Integrated Agriculture-Aquaculture Farming Systems, ICLARM Conference Proceedings 4. ICLARM-SEARCA, Manila, Philippines.
7. Aquaculture Information Center, National Agricultural Library, Room, 304, Beltsville, MD 20705, 301–344–3704

Biography

Lora Petrie-Hanson is an Associate Professor at the College of Veterinary Medicine, Mississippi State University, where she has worked as a fish health researcher/Fish Disease Diagnostician/Aquatic Facilities Manager since 1990. She obtained her BS (Fisheries Science, 1981) and MS (Fisheries-hatchery science, 1984) degrees at Auburn University, and her Ph.D. in Veterinary Medical Sciences from the College of Veterinary Medicine, Mississippi State University in 1997. She has worked with the spawning and rearing of warm water aquaculture species since 1981 and supervises the production and rearing of all fish used in aquatic medicine research at CVM-MSU. She is an American Fisheries Society Fish Health Section Certified Fish Pathologist. Her research focuses on health problems in fry production systems and functional immunity of developing fish.

Larry Hanson is a Professor at the College of Veterinary Medicine, Mississippi State University where he has been on the faculty since 1990. He obtained his BS (Fisheries Science, 1981) and MS (Fisheries-Fish Pathology, 1983) degrees at Auburn University and his Ph.D. in Veterinary Medical Sciences from the School of Veterinary Medicine, Louisiana State University. He is an American Fisheries Society Fish Health Section Certified Fish Pathologist and has served as a fish disease diagnostician working with aquaculture industries since 1983. His research program focuses on the molecular basis of fish diseases and fish vaccine development.

Lydia Brown is a British Veterinarian who obtained her Ph.D. in Aquatic Veterinary Studies from the Institute of Aquaculture in Scotland in 1983. She has worked with fish throughout Northern Europe, Israel and Hong Kong. While doing a Visiting Scientist Residency at Mississippi State University she helped with the original edition of Raising Healthy Fish.

This book was financed in part by
Christian Veterinary Mission members from Mississippi.

NOTES

NOTES

Made in the USA
Monee, IL
04 November 2026

67379247R00056